P9-DGL-222

SCHAUM'S *Easy* OUTLINES

# HUMAN ANATOMY AND PHYSIOLOGY

## Other Books in Schaum's Easy Outline Series Include:

Schaum's Easy Outline: Calculus
Schaum's Easy Outline: College Algebra
Schaum's Easy Outline: College Mathematics
Schaum's Easy Outline: Geometry
Schaum's Easy Outline: Mathematical Handbook of Formulas and Tables
Schaum's Easy Outline: Statistics
Schaum's Easy Outline: Principles of Accounting
Schaum's Easy Outline: Biology
Schaum's Easy Outline: College Chemistry
Schaum's Easy Outline: Organic Chemistry
Schaum's Easy Outline: Physics
Schaum's Easy Outline: Programming with C++
Schaum's Easy Outline: Programming with Java
Schaum's Easy Outline: French
Schaum's Easy Outline: German
Schaum's Easy Outline: Spanish
Schaum's Easy Outline: Writing and Grammar

SCHAUM'S *Easy* OUTLINES

# HUMAN ANATOMY AND PHYSIOLOGY

BASED ON SCHAUM'S *Outline of Theory and Problems of Human Anatomy and Physiology* BY KENT M. VAN DE GRAAFF AND R. WARD RHEES

ABRIDGEMENT EDITOR
PATRICIA BRADY WILHELM

SCHAUM'S OUTLINE SERIES
McGRAW-HILL

*New York Chicago San Francisco Lisbon London Madrid Mexico City Milan New Delhi San Juan Seoul Singapore Sydney Toronto*

**KENT M. VAN DE GRAAFF** is Professor of Zoology at Weber State University in Ogden, Utah, where he received his B.S. in zoology. He holds an M.S. from the University of Utah and a Ph.D. from Northern Arizona University. The author or co-author of several college textbooks, he previously taught anatomy at the University of Minnesota and Brigham Young University.

**R. WARD RHEES** is Professor of Zoology at Brigham Young University. He received his B.S. in pharmacy from the University of Utah and a Ph.D. in physiology from Colorado State University. He previously taught at Weber State University and was a visiting professor at the Department of Anatomy and the Brain Research Institute at the UCLA School of Medicine. He has published articles based on his research in leading scholarly journals.

**PATRICIA BRADY WILHELM** teaches human anatomy and zoology at the Community College of Rhode Island in Warwick. She holds a B.A. from Cornell University and a Ph.D. from Brown University, where she also previously taught comparative vertebrate anatomy and ecology. She has been a consultant to the teaching with technology program at the University of Rhode Island and has helped develop an anatomy on-line Web site.

Copyright © 2001 by The McGraw-Hill Companies, Inc. All rights reserved. Printed in the United States of America. Except as permitted under the Copyright Act of 1976, no part of this publication may be reproduced or distributed in any form or by any means, or stored in a data base or retrieval system, without prior written permission of the publisher.

1 2 3 4 5 6 7 8 9 0 DOC DOC 0 9 8 7 6 5 4 3 2 1 0

ISBN 0-07-136976-7

*McGraw-Hill*

*A Division of The McGraw-Hill Companies*

# Contents

# Chapter 1
# Introduction to the Human Body

In This Chapter:

- ✔ *Humans as Biological Organisms*
- ✔ *Levels of Organization of the Human Body*
- ✔ *Homeostasis*
- ✔ *Anatomical Position and Terminology*
- ✔ *Body Regions and Body Cavities*
- ✔ *Solved Problems*

Anatomy and Physiology are subdivisions of the science of biology, the study of living organisms. Human anatomy is the study of body structure and the relationships between body structures. Human physiology is concerned with the functions of the body parts. In general, function is determined by structure.

## Humans as Biological Organisms

Human beings (*Homo sapiens*) are biological organisms. The basic physical requirements of humans, as with all organisms, are: *water*, for a variety of metabolic processes; *food*, to supply energy, raw materials for building new living matter, and chemicals necessary for vital reactions; *oxygen*, to release energy from food materials; *heat*, to promote chemical reactions; and *pressure*, to allow breathing.

## Levels of Organization of the Human Body

The levels of organization of the human body are, from the simplest to the most complex: chemical, cellular, tissue, organ, system, and organism. Each level of body organization represents an association of units from the preceding level.

**Chemical** and **cellular** levels are the basic structural and functional levels.

A **tissue** is an aggregation of similar cells that performs a specific function. There are four types of tissues found in humans.

### Types of Tissues

1. **Epithelial tissue:** Covers body and organ surfaces, lines body cavities, and forms glands. Is involved with protection, absorption, excretion, secretion, diffusion, and filtration.
2. **Connective tissue** binds, supports, and protects body parts, stores energy and minerals.
3. **Muscle tissue** contracts to produce movement.
4. **Nervous tissue** initiates and transmits nerve impulses that coordinate body activities.

An **organ** is composed of several tissue types that are integrated to perform a particular function.

A **system** is an organization of two or more organs and associated tissues working as a unit to perform a common function or set of functions. The body systems are:

1. The **muscular and skeletal systems** function in body support and locomotion.
2. The **endocrine and nervous systems** function in integration and coordination by maintaining consistency of body functioning.
3. The **digestive, respiratory, circulatory, lymphatic, and urinary systems** are involved with processing and transporting body substances. The **digestive system** mechanically and chemically breaks down foods for cellular use and eliminates undigested wastes. The **respiratory system** supplies oxygen to the blood, eliminates carbon dioxide, and helps regulate acid-base balance. The **circulatory system** transports respiratory gases, nutrients, wastes, and hormones; helps regulate body temperature and acid-base balance; and protects against disease and fluid loss. The **lymphatic system** transports lymph from tissues to the blood stream, defends the body against infections, and aids in the absorption of fats. The **urinary system** functions to remove wastes from the blood; regulate the chemical composition, volume, and electrolyte balance of the blood; and helps to maintain the acid-base balance of the body.
4. The **integumentary system** functions to protect the body, regulate body temperature, eliminate wastes, and receive sensory stimuli.
5. The **reproductive system** functions to produce gametes for sexual reproduction and to produce sex hormones.

## Homeostasis

Homeostasis is the process by which a nearly stable internal environment is maintained in the body so that cellular metabolic functions can proceed at maximum efficiency. Homeostasis is maintained by muscles or glands that are regulated by sensory information from the internal environment.

## Anatomical Position and Terminology

All terms of direction that describe the relationship of one body part to another are made in reference to a standard anatomical position. In **anatomical position,** the body is erect, feet are parallel and flat on the floor, eyes are directed forward, and arms are at the sides of the body with the palms of the hands turned forward and the fingers pointing downward.

Descriptive and directional terms are used to communicate the position of structures, surfaces, and regions of the body with respect to anatomical position. Commonly used terms are listed and defined in Table 1.2.

In addition to the terms listed in Table 1.2, three planes of reference are used to locate and describe structures within the body. The **midsagittal plane** is the plane of symmetry, dividing the body into right and left halves. A **coronal plane** divides the body into front and back portions, and a **transverse (horizontal or cross-sectional) plane** divides the body into superior and inferior portions.

**Table 1.2** Commonly Used Anatomical Descriptive and Directional Terms

| *Term* | *Definition* |
|---|---|
| Superior (cranial) | Toward the head |
| Inferior (caudal) | Toward the bottom (tail) |
| Anterior (ventral) | Toward the front |
| Posterior (dorsal) | Toward the back |
| Medial | Toward the midline of the body |
| Lateral | Toward the side of the body |
| Internal (deep) | Away from the surface of the body |
| External (superficial) | Toward the surface of the body |
| Proximal | Toward the main mass of the body |
| Distal | Away from the main mass of the body |
| Visceral | Related to internal organs |
| Parietal | Related to the body walls |

## Body Regions and Body Cavities

The principal **body regions** are the *head, neck, trunk* (divided into the thorax and the abdomen), *upper extremity* (two), and *lower extremity* (two).

The **body cavities** are confined spaces in which organs are protected, separated, and supported by associated membranes. The **posterior (dorsal) cavity** includes the **cranial** and **vertebral cavities** and contains the brain and spinal cord.

The **anterior (ventral) cavity** includes the **thoracic, abdominal,** and **pelvic cavities** and contains the visceral organs. The abdominal cavity and the pelvic cavity are frequently referred to collectively as the **abdominopelvic cavity** because there is no physical division between these two regions. The visceral organs located in the thoracic cavity are the heart and lungs. The thoracic cavity is partitioned into two **pleural cavities,** one surrounding each lung and the **pericardial cavity** surrounding the heart. The area between the two lungs is known as the **mediastinum.** The viscera of the abdominal cavity include the stomach, small intestine, large intestine, spleen, liver, and gallbladder.

The body cavities serve to segregate organs and systems by function: the major portion of the nervous system occupies the posterior cavity; the principal organs of the respiratory and circulatory systems are in the thoracic cavity; the primary organs of digestion are in the abdominal cavity; and the reproductive organs are in the pelvic cavity.

Body membranes, composed of thin layers of connective and epithelial tissue, serve to cover, protect, lubricate, separate, or support visceral organs or to line body cavities. The two principal types are **mucous membranes** and **serous membranes.**

Mucous membranes secrete a thick, viscous substance called mucous that lubricates and protects the body organs where it is secreted. Examples of mucous membranes are the epithelial membranes lining the nasal cavity, the trachea, and the oral cavity. Mucous membranes are found lining the inside walls of many other body organs.

Serous membranes line the thoracic and abdominopelvic cavities and cover the visceral organs (described above). They are composed of thin sheets of epithelial tissue that lubricate, support, and compartmentalize visceral organs. **Serous fluid** is the watery lubricant they secrete. The serous membranes of the thoracic cavity are the **parietal** and **visceral pleura,** lining the thoracic walls and diaphragm and the outer surface of the lungs respectively, and the **parietal** and **visceral pericardium** surrounding the heart. The serous membranes of the abdominopelvic cavity are the **parietal** and **visceral peritoneum,** lining the abdominal wall and covering the abdominal viscera respectively; and the **mesentery**, a double fold of which supports the viscera and loosely anchors it to the abdominal body wall.

## You Need to Know

- The levels of organization of the human body.
- The body systems and their function.
- The definition of homeostasis.
- Standard anatomical position.
- The meaning of the anatomical terms.
- The body cavities and the membranes surrounding them.

## Solved Problems

### True or False

____ 1. A group of cells cooperating in a particular function is called a tissue. (**True**)

____ 2. The term *parietal* refers to the body wall, and the term *visceral* refers to internal body organs. (**True**)

____ 3. The thumb is lateral to the other digits of the hand and distal to the antebrachium. (**True**)

____ 4. In anatomical position the subject is standing erect, the feet are together, and the arms are relaxed to the side of the body with the thumbs forward. (**False**)

____ 5. Mesenteries tightly bind visceral organs to the body wall so that they are protected from excessive movement. (**False**)

____ 6. Increased body temperature during exercise is an example of a homeostatic feedback mechanism. (**False**)

# Chapter 2
# Cellular Chemistry

In This Chapter:

- ✔ *Atoms*
- ✔ *Molecules and Chemical Bonds*
- ✔ *Solutions and Properties of Solutions*
- ✔ *Organic and Inorganic Compounds*
- ✔ *Solved Problems*

All matter, living and nonliving, consists of building units called *chemical elements*. Ninety-six percent of the human body is composed of the chemicals Carbon (C), Nitrogen (N), Oxygen (O), and Hydrogen (H). Calcium (Ca), Phosphorus (P), Potassium (K), and Sulfur (S) make up 3 percent of the body. The remainder of the body is composed of small quantities of Iron (Fe), Chlorine (Cl), Iodine (I), Sodium (Na), Magnesium (Mg), Copper (Cu), Manganese (Mn), Cobalt (O), Zinc (Zn), Chromium (Cr), Florine (F), Molybdenum (Mo), Silicon (Si), and Tin (Sn) referred to as trace elements.

## Atoms

An **atom** is the smallest unit of an element that retains its chemical properties. Every pure element is composed of only one kind of atom.

An atom is composed of three kinds of elementary particles:

- **protons:** Charge of +1, mass of 1.
- **neutrons:** No electrical charge, mass of 1.
- **electrons:** Charge of -1, very small mass.

Protons and neutrons are bound in the nucleus of the atom.

The **atomic number (Z)** = The number of protons in the nucleus.
The **atomic mass** = The number of protons + the number of neutrons (or 2 times the atomic number).

The atomic number is the same for all atoms of a given chemical element. Surrounding the nucleus are precisely Z electrons, making the atom as a whole electrically neutral. Electrons orbit the nucleus, much as the planets of the solar system orbit the sun. The distribution of electrons is organized into energy levels (shells). The electrons are distributed among the shells. The capacities of the first four shells are 2, 8, 8, and 18 electrons. The atom is built by one electron at a time, with a given shell entered only if all interior shells are full.

## Molecules and Chemical Bonds

A **molecule** is a combination of two or more atoms, joined by *chemical bonds*. Just as atoms are the smallest units of a chemical element, molecules are the smallest unit of a chemical compound. Water is a chemical compound that is essential for life. It consists of molecules, each containing one oxygen atom and two hydrogen atoms ($H_2O$). In chemical notation, subscripts denote how many atoms of each element are in one molecule of the compound.

Molecules are held together by attractive forces called **bonds. Ionic bonds** form when atoms give up or gain electrons and become either positively or negatively charged. The charged atoms are called

**ions,** and those with negative charges are attracted strongly to those with positive charges (Figure 2-1). **Covalent bonds** form when atoms share electrons (Figure 2-2). A **hydrogen bond** is a weak bond between molecules that forms when hydrogen forms a covalent bond with another atom, or example oxygen. The hydrogen atom gains a slight positive charge and has an affinity for the slightly negatively charged oxygen of other molecules. *Chemical reactions* occur when molecules form, are broken, or rearrange their component atoms.

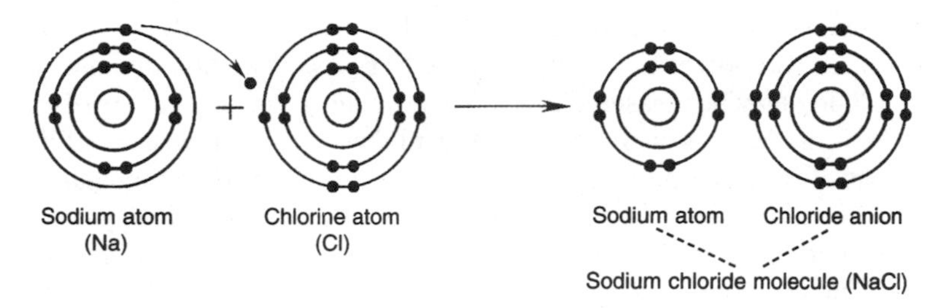


**Figure 2-1.** The formation of an **ionic bond** in the NaCl molecule.

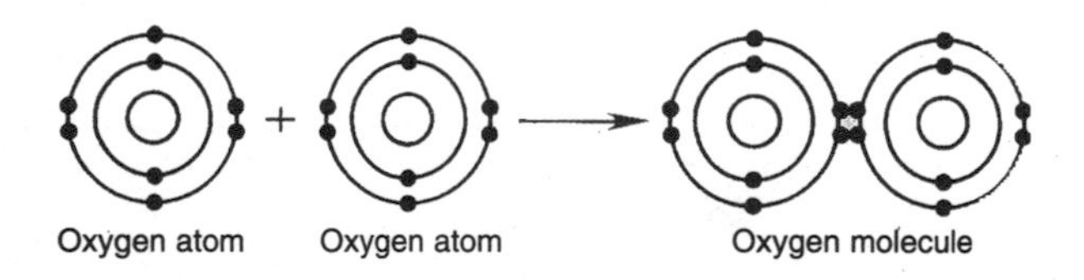


**Figure 2-2.** The formation of a **covalent bond** in the $O_2$ molecule.

## Remember

### Types of Chemical Bonds

Ionic bonds
Covalent bonds
Hydrogen bonds

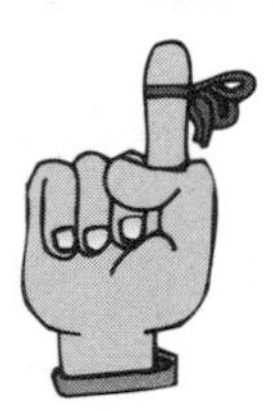

Many of the unique properties of water, freezing and boiling points, surface tension, adhesion, cohesion, and capillary action are due to the hydrogen bonding between water molecules.

## Solutions and Properties of Solutions

When two or more substances combine without forming bonds with each other, the result is a **mixture. Solutions** are mixtures in which the molecules of all the combined substances are distributed homogeneously throughout the mixture. Solutions include solids (**the solute**) dissolved in liquid (**the solvent**) as with salt water. The concentration of solute in a solution may be measured in many ways such as the percentage of the solutes in the solution, or the **molarity** of the solution, a measure of the moles of solute per liter of solution (1 mole = 6.022 × 1023 molecules).

One important property of solutions is the acidity or basicity (alkalinity) of the solution. This property is measured by the **pH** of the solution. In any sample of water, a certain proportion of water molecules exits in an ionized form as H+ (hydrogen ions) and OH- (hydroxide ions). In pure water the number of H+ equals the number of OH-, and the solution is referred to as neutral, with a pH of 7.

An **acid** is a substance that when added to water increases the concentration of H+ ions. The **pH of an acid is less than 7.**

A **base** is a substance that when added to water increases the concentration of OH- ions. **The pH of a base is greater than 7 up to 14.**

The lower the pH number, the greater the acidity, the higher the number, and the greater the alkalinity of the solution.

A **salt** is an ionic compound formed from the residue of an acid and the residue of a base. When an acid loses its proton (H+) and a base loses a hydroxyl group (OH-) the remaining ions of the molecules will sometimes bind together forming a salt. For example, the formation of table salt is represented below:

$$\underset{\text{acid}}{HCL} + \underset{\text{base}}{NaOH} \rightarrow \underset{\text{water}}{H_2O} + \underset{\text{salt}}{NaCl}$$

A **buffer** is a combination of a weak acid and its salt in a solution that has the effect of stabilizing the pH of the solution. If a solution contains a buffer, its pH will not change dramatically even when strong acids or bases are added. When acid is added to the solution, it is neutralized by the salt of the weak acid. When a base is added to the solution it is neutralized by the weak acid itself. Three important buffer systems found in the body are the **bicarbonate buffer** found in the blood and the extracellular fluid, the **phosphate buffer** in the kidneys and the intracellular fluid, and the **protein buffer** found in all tissues.

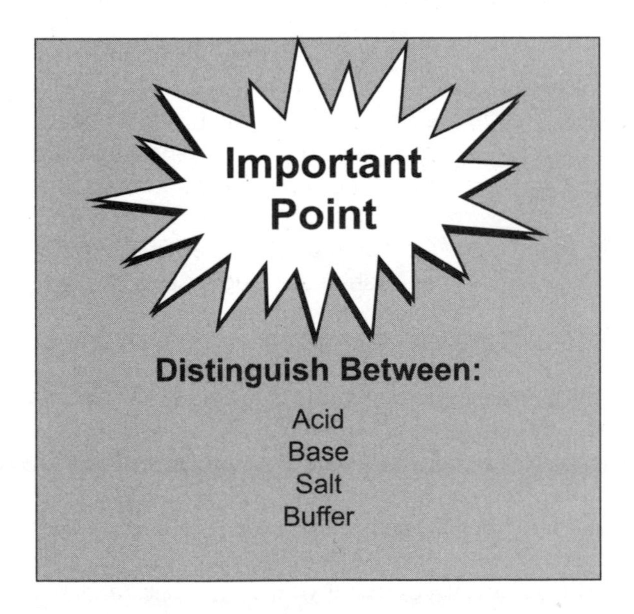


## Organic and Inorganic Compounds

**Inorganic compounds** do not contain carbon (exceptions include CO and $CO_2$) and are usually small molecules. **Organic compounds**

always contain carbon and are held together by covalent bonds. Organic compounds are usually large, complex molecules. Both inorganic and organic compounds are important in the chemical processes that are essential to life.

Some inorganic compounds important in living organisms are water, oxygen, carbon dioxide, salts, acids, bases, and electrolytes such as sodium (Na+), potassium (K+), calcium (Ca2+), and chloride (Cl-). These electrolytes are important in the transmission of nerve impulses, maintenance of body fluids, and functioning of enzymes and hormones.

## Four Families of Organic Compounds

| | |
|---|---|
| Carbohydrates | Composed of carbon, oxygen, and hydrogen. Classified as monosaccharides, disaccharides, and polysaccharides. |
| Lipids | Composed of fatty acids and glycerol. |
| Proteins | Composed of amino acids. The function of a protein is determined by the character of the amino acids it contains. |
| Nucleic acids | Composed of nucleotides containing a phosphate, a sugar, and a nitrogenous base. |

The nucleic acids **deoxyribonucleic acid (DNA)** and **ribonucleic acid (RNA)** are composed of **nucleotides.** A nucleotide has three parts: a phosphate group, a pentose sugar, and a nitrogenous base. The pentose sugar is always ribose in RNA and deoxyribose in DNA. The phosphate remains constant from one nucleotide to the next, but the base (in DNA) may be either adenine (A), thymine (T), guanine (G), or cytosine (C). RNA substitutes uracil (U) for thymine. The nucleotides are joined together into the macromolecules DNA and RNA.

## Solved Problems

1. A neutral atom contains (*a*) the same number of electrons as it does protons, (*b*) more protons than electrons, (*c*) the same number of electrons as it does neutrons, (*d*) more electrons than protons. (**a**)

2. Bonds that result from transfer of electrons are called (*a*) ionic bonds, (*b*) covalent bonds, (*c*) peptide bonds, (*d*) polar bonds, (*e*) all of above. (**a**)

3. Of the following nitrogenous bases, which is found exclusively in RNA? (*a*) thymine, (*b*) guanine, (*c*) adenine, (*d*) uracil. (**d**)

4. Which of the following is a false statement? (*a*) Acids increase hydrogen ion concentration in solution. (*b*) Acids act as proton donors. (*c*) Acids yield a higher hydroxide (OH-) concentration than a hydrogen ion concentration. (*d*) Acids have a low pH. (**c**)

# Chapter 3
# Cell Structure and Function

In This Chapter:

- ✔ *Cell Structures*
- ✔ *Replication, Transcription, and Translation*
- ✔ *Mitosis and Meiosis*
- ✔ *Cellular Communication*
- ✔ *Solved Problems*

**Prokaryotic cells** lack a membrane-bound nucleus; instead they contain a single strand of nucleic acid. These cells contain few organelles. A rigid or semirigid cell wall surrounding a cell (plasma) membrane gives the cell its shape. Bacteria are prokaryotic single-celled organisms.

**Eukaryotic cells** contain a true nucleus with multiple chromosomes. They also have several types of specialized membrane-bound organelles. Like prokaryotes, eukaryotic cells have a cell (plasma) membrane. Since all human cells are eukaryotic, most of this chapter will focus on eukaryotic cells and their functions. Organisms composed of eukaryotic cells include protozoa, fungi, algae, plants, invertebrates, and vertebrate animals.

## Cell Structures

The following cell structures are common to all cells:

**Cell (plasma) membrane.** Both prokaryotic and eukaryotic cells are bound by a cell (plasma) membrane. The cell membrane is selectively permeable, allowing certain substances to pass through, but not others. Water, alcohol, and gases readily pass through the cell membrane, but ions, large proteins, and carbohydrates do not. Substances can pass through the cell membrane via one of the processes listed below.

| | |
|---|---|
| **Diffusion** | Passive movement of any substance from an area of high concentration to an area of low concentration. |
| **Osmosis** | Passive diffusion of water through a semipermeable membrane. |
| **Facilitated transport** | Accomplished by proteins in the membrane that allow the passage of otherwise restricted molecules. |
| **Active transport** | The process of using energy (ATP) to "pump" molecules across the membrane against the concentration gradient. |
| **Endocytosis**<br>**Phagocytosis**<br>**Pinocytosis** | Processes that bring material into the cell.<br>Cell membrane engulfs a foreign substance or body. Membrane engulfs small droplets of water. |
| **Exocytosis** | Release of molecules from the cell. |

**Cytoplasm.** Cytoplasm is the fluid matrix within a cell. It consists primarily of water and dissolved substances including $O_2$, $CO_2$, cellular wastes (urea), glucose, ions, proteins, and ATP.

**Ribosomes.** Ribosomes are the protein factories of the cell. They are responsible for translation, building proteins as directed by mRNA synthesized by the DNA in the nucleus.

**Cell wall** (found in some prokaryotic and eukaryotic cells). A cell wall is present in only eukaryotic plant cells. It functions to maintain cell shape. Some prokaryotic bacteria also have cell walls.

**Chromosomes.** The chromosomes of a cell contain the DNA. In eukaryotic cells, the chromosomes are located within the nucleus. Every species has a distinct number of chromosomes.

An **organelle** is any subcellular structure having a specific function. The organelles described below are found only in eukaryotic cells.

| Organelle | Structure and Function |
|---|---|
| **Nucleus** | Contains chromosomes and the nucleolus. Stores genetic material and controls all cellular activities. |
| **Nucleolus** | Mass of RNA located within the nucleus. Center for organizing ribosomes and other products with RNA. |
| **Ribosomes** | Granular protein particles involved with the synthesis of proteins. |
| **Endoplasmic reticulum (ER)** | Membranous network continuous with the cell and nuclear membranes. |
| **Rough ER** | ER with ribosomes attached. Involved with protein synthesis for use outside the cell. |
| **Smooth ER** | No ribosomes attached. Functions in steroid synthesis, intercellular transport, and detoxification. |

| Organelle | Structure and Function |
|---|---|
| **Golgi apparatus** | Stacked membrane and vessels that package proteins produced at rough ER. |
| **Mitochondria** | Oval organelles with membrane folds called cristae that produce ATP through the Krebs cycle and oxidative phosphorylation. |
| **Lysosomes** | Vesicles filled with enzymes that break-down worn cellular components or engulfed particles. |
| **Secretory vesicles** | Membrane-bound sacs which store proteins for secretion. |
| **Microtubules Microfilaments** | Long protein fibers that function in cell support and movement. |
| **Centrioles** | Two short rods composed of microtubules located near the nucleus division that are involved in movement of chromosomes during cell microtubules. |

## Replication, Transcription, and Translation

**Replication** refers to the process in which DNA makes an identical copy of itself prior to cell division. **Transcription** refers to making mRNA from the DNA template. The mRNA then leaves the nucleus and joins with a ribosome in the cytoplasm to synthesize a protein in a process called **translation.** A mutation is an unrepaired mistake in replication. Some occur spontaneously, but many are induced by various substances called mutagens. Most mutations are harmless or unnoticeable, some may be harmful or lethal, other may be beneficial. Mutations play an important role in creating diversity in the genetic makeup of a species.

## Mitosis and Meiosis

**Mitosis** is the process of normal cell division (Figure 3-1). It occurs whenever the body cells need to produce more cells for growth or for replacement and repair. The result of mitosis is two identical *daughter cells* with the same chromosomal content as the parent cell.

**Be Certain You Can Distinguish Between:**

**Mitosis:** Cell division, number of chromosomes unchanged.

**Meiosis:** Cell division that reduces the number of chromosomes in daughter cells by one half. Occurs only in gamete formation.

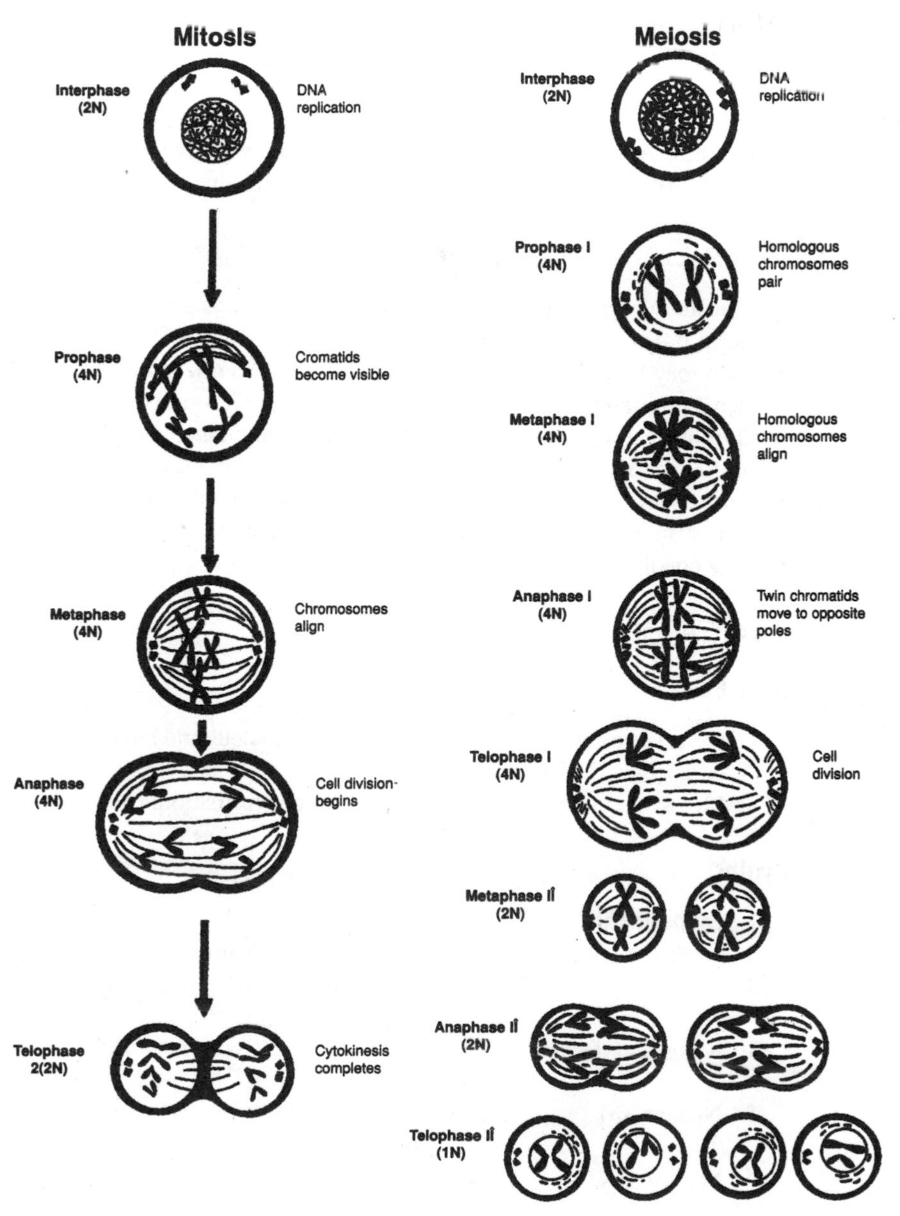


**Figure 3-1.** Stages of mitosis and meiosis.

**Meiosis** is the process of gamete (sex cell) formation. It resembles mitosis in many ways, except that the end result is four daughter cells, each with half the chromosomal content of the parent cell (Figure 3-1). In humans, the somatic (body) cells all have 46 chromosomes or 23 pairs (diploid). After undergoing meiosis, the gametes have only 23 chromosomes (haploid) or half the number of chromosomes.

## Cellular Communication

Cells adjacent to each other or distant from each other must often communicate in order for a body system to function normally. This communication may be accomplished in several ways. **Chemical messengers,** such as hormones, secreted into the blood, or neurotransmitters, released from a nerve cell, can accelerate or inhibit cellular functioning. Physical contact of one cell with another may trigger **contact inhibition,** which is frequently expressed as a repression of mitosis. When cells fail to respond to contact inhibition and continue to divide uncontrolled, the condition is called **cancer**.

## Solved Problems

### True or False

____ 1. Eukaryotic cells lack a membrane-bound nucleus and have few organelles. (**False**)

____ 2. Meiosis is the process of gamete (sex cell) formation. (**True**)

### Matching

| | |
|---|---|
| ____ 1. Lysosome (**b**) | (a) control center of cell |
| ____ 2. Centriole (**d**) | (b) vesicle containing hydrolytic enzymes |
| ____ 3. Golgi apparatus (**e**) | (c) synthesis of steroids and detoxification. |
| ____ 4. Ribosome (**f**) | (d) movement of chromosomes during mitosis |
| ____ 5. Nucleus (**a**) | (e) formation of secretory vesicles and lysosomes |
| ____ 6. Smooth ER (**c**) | (f) synthesis of proteins |

# Chapter 4
# TISSUES

IN THIS CHAPTER:

- ✔ *Epithelial Tissue*
- ✔ *Connective Tissue*
- ✔ *Muscle Tissue*
- ✔ *Nervous Tissue*
- ✔ *Solved Problems*

A tissue is an aggregation of similar cells that perform a specific set of functions. The body is composed of over 25 kinds of tissues, classified as **epithelial tissue, connective tissue, muscle tissue,** and **nervous tissue.**

## Epithelial Tissue

Epithelial tissue covers body and organ surfaces, lines body cavities and lumina, and forms various glands. It functions in protection, absorption, excretion, and secretion. The outer surface of epithelium is exposed either to the outside of the body or to a lumen or cavity within the body. The deep inner surface is bound by a basement membrane. Epithelial tissue is avascular (without blood vessels) and composed of tightly packed cells. Epithelium is classified by

1. the number of layers of cells: a single layer is referred to as **simple** epithelium, multilayered epithelium is called **stratified;** and
2. by the shape of the cells: **squamous** (flat), **cuboidal,** or **columnar**.

The stratified squamous epithelium of the epidermal layer of the skin contains the protein **keratin,** which functions to waterproof the skin. **Transitional epithelium** is similar to nonkeratinized stratified squamous epithelium, except that the surface cells of the former are large and round rather than flat, and they may have two nuclei. Transitional epithelium is specialized to permit distension of the ureters and urinary bladder.

During development, certain epithelial cells invade the underlying connective tissue and form specialized secretory accumulations called glands. **Exocrine glands** retain a connection to the surface in the form of a duct. The three types of exocrine glands are **merocrine, apocrine,** and **holocrine** glands. **Endocrine glands** lack ducts and secrete their products (hormones) directly into the bloodstream.

**Table 4.1** Classification of Epithelial Tissue

| Type | Structure and Function | Location |
|---|---|---|
| Simple squamous epithelium | Single layer of flattened cells; diffusion and filtration | Forming capillary walls; lining air sacs (alveoli) of lungs; covering visceral organs; lining body cavities |
| Simple cuboidal epithelium | Single layer of cube-shaped cells; excretion, secretion, or absorption | Covering surface of ovaries; lining kidney tubules, salivary ducts, and pancreatic ducts |
| Simple columnar epithelium | Single layer of nonciliated column-shaped cells; protection, secretion and absorption | Lining digestive tract, gall bladder, and excretory ducts of some glands |
| Simple ciliated columnar epithelium | Single layer of ciliated column-shaped cells; transport role through ciliary motion | Lining uterine (fallopian) tubes and limited areas of respiratory tract |
| Pseudo-stratified ciliated columnar epithelium | Single layer of ciliated, irregularly shaped cells; protection, secretion, ciliary motion | Lining respiratory passageways and auditory tubes |
| Stratified squamous epithelium keratinized | Multilayered, contains keratin, outer layers flattened and dead; protection | Epidermis of the skin |
| Stratified squamous epithelium (non-keratinized) | Multilayered, lacks keratin, outer layers moistened and alive; protection and pliability | Linings of oral and nasal cavities, esophagus, vagina, and anal canal |
| Stratified cuboidal epithelium | Usually two layers of cube-shaped cells; strengthening of luminal walls | Ducts of larger sweat glands, salivary glands, and pancreas |
| Transitional epithelium | Numerous layers of rounded nonkeratinized cells; distension | Lining urinary bladder and portions of ureters and urethra |

## Connective Tissue

One of the most important components of connective tissue is the **matrix**, a bed of secreted organic material of varying composition that binds widely separated cells of a tissue. Connective tissue supports and binds other tissues, stores nutrients, and/or manufactures protective and regulatory materials.

**Table 4.2** Classification of Connective Tissue

| Type | Cells and Matrix | Function | Location |
|---|---|---|---|
| Loose (areolar) | Fibroblasts, mast cells; Collagenous fibers, elastin | Binding; protection, nourishment; holds fluids | Deep to skin; around muscles, vessels, organs |
| Dense fibrous | Fibroblasts; Densely packed collagenous fibers | Strong, flexible | Tendon, ligament |
| Elastic | Fibroblasts; Elastin fibers | Flexibility, distensibility | Arteries, larynx, trachea, bronchi |
| Reticular | Phagocytes; Reticular fibers in jellylike matrix | Phagocytic function | Liver, spleen, lymph nodes, bone marrow |
| Adipose | Adipocytes; very little matrix | Stores lipids | Hypodermis, around organs |
| Cartilage<br>Hyaline<br>Fibrocartilage<br>Elastic | Chondrocytes; Collagenous fibers, elastin in elastic cartilage | Support, strength and flexibility | Joints, trachea, nose, outer ear, larynx |
| Bone<br>Spongy bone<br>Compact bone | Osteocytes; Collagenous fibers, calcium carbonate | Strong support | Bones |
| Blood | Erythrocytes; leukocytes, thrombocytes (platelets); Plasma | Conduction of nutrients and wastes | Circulatory system |

## Muscle Tissue

By contracting, muscle tissue moves materials through the body, enables movement of one part of the body with respect to another, and allows locomotion. Muscle cells, also called muscle fibers, are elongated in the direction of contraction, and movement is accomplished through the shortening of the fibers in response to a stimulus. In addition to the contractile properties of muscle, all muscle fibers are irritable, responding to nervous stimuli, extendible and elastic. There are three types of muscle tissue in the body: *smooth, cardiac,* and *skeletal.*

**Table 4.3** Features of Muscle Tissue

| Type | Location | Structure and Function |
| --- | --- | --- |
| Smooth muscle | Walls of hollow organs | Elongated, spindle-shaped fiber with single nucleus; non-striated, involuntary |
| Cardiac muscle | Wall of the heart | Branched, striated fiber with single nucleus and intercalated discs; involuntary rhythmic contraction |
| Skeletal muscle | Spanning joints of skeleton via tendons | Multinucleated, striated, cylindrical fibers; voluntary |

Metabolism within cells releases heat as an end product. Muscles account for nearly one-half of the body weight, and even the fibers of resting muscles are in a continuous state of fiber activity (tonus). Thus, muscles are major heat sources. Maintaining a high body temperature is of homeostatic value in providing optimal conditions for metabolism.

## Nervous Tissue

Nervous tissue consists of mainly two types of cells: neurons and neuroglia. **Neurons,** or nerve cells, are highly specialized to conduct impulses, called *action potentials*. **Neuroglia** primarily function to support and assist neurons. Neuroglia are about five times as abundant as neurons and they have mitotic capabilities throughout life. Neurons have branched **dendrites** that extend from the surface of the cell body to provide a large surface area for receiving stimuli and conducting impulses to the cell body. The elongated **axon** conducts the impulse away from the cell body to another neuron or to an organ that responds to the impulse. There are six types of neuroglia. Four are found in the central nervous system (CNS): astocytes, ependymal cells, oligodentrocytes, and microglia. The remaining two, ganglionic gliocytes (satellite cells) and neurolemmocytes (Schwann cells) are located in the peripheral nervous system (PNS). **Neurolemmocytes** support the axon by ensheathing it with a lipid-protein substance, **myelin.** This myelin sheath aids in the conduction of nerve impulses and promotes regeneration of a damaged neuron.

**Remember**

There are only four major tissues in the human body:

- Epithelial tissue
- Connective tissue
- Muscle tissue
- Nervous tissue

Each major tissue has multiple distinctive tissue types. For each tissue type you should know the:

- Structure of each tissue
- Function of each tissue
- Location of each tissue

## Solved Problems

1. Epithelium consisting of two or more layers is classified as ________________. (**stratified epithelium**)
2. Rhythmic contractions of sheets of ______________ muscle tissue in the intestinal wall results in involuntary movement of food materials. (**smooth**)
3. ______________ is the matrix of blood tissue. (**Plasma**)

4. The ____________ of a neuron receive a stimulus and conduct the nerve impulse to the cell body. (**dendrites**)
5. ____________ muscle tissue is composed of multinucleated, striated, cylindrical fibers arranged into fasciculi. (**Skeletal**)
6. The lipid-protein product of neurolemmocytes (Schwann cells) forms a cover of ____________ around the axon of a neuron. (**myelin**)

# Chapter 5
# INTEGUMENTARY SYSTEM

IN THIS CHAPTER:

- ✔ *Functions of the Integumentary System*
- ✔ *Structure of the Skin*
- ✔ *Associated Structures of the Skin*
- ✔ *Physiology of the Skin*
- ✔ *Solved Problems*

The integumentary system is composed of the **skin,** or **integument,** and associated structures (hair, glands, and nails). This system accounts for approximately 7 percent of the body weight and is a dynamic interface between the body and the external environment.

## Functions of the Integumentary System

The functions of the integumentary system include physical protection, hydroregulation, thermoregulation, cutaneous absorption, synthesis, sensory reception, and communication. The skin is a physical barrier to most microorganisms, water, and most UV light. The acidic surface (pH 4.0–6.8) retards the growth of most pathogens. The skin protects the body

from desiccation (dehydration) when on dry land and from water absorption when immersed in water. A normal body temperature of 37°C (98.6 °F) is maintained by the antagonistic effects of shivering and sweating, as well as by vasodilation and vasoconstriction of the blood vessels to the skin. The skin permits the absorption of small amounts of UV light necessary for synthesis of vitamin D. It is important to note that certain toxins and pesticides also may enter the body through cutaneous absorption. The skin synthesizes **melanin** (a protective pigment) and **keratin** (a protective protein). Numerous sensory receptors are located in the skin, especially in parts of the face, palms, and fingers of the hands, soles of the feet, and genitalia. The skin interacts with numerous body systems in accomplishing these various functions including the circulatory system, the immune system, and the nervous system.

## Structure of the Skin

A diagram of the skin is shown in Figure 5-1.

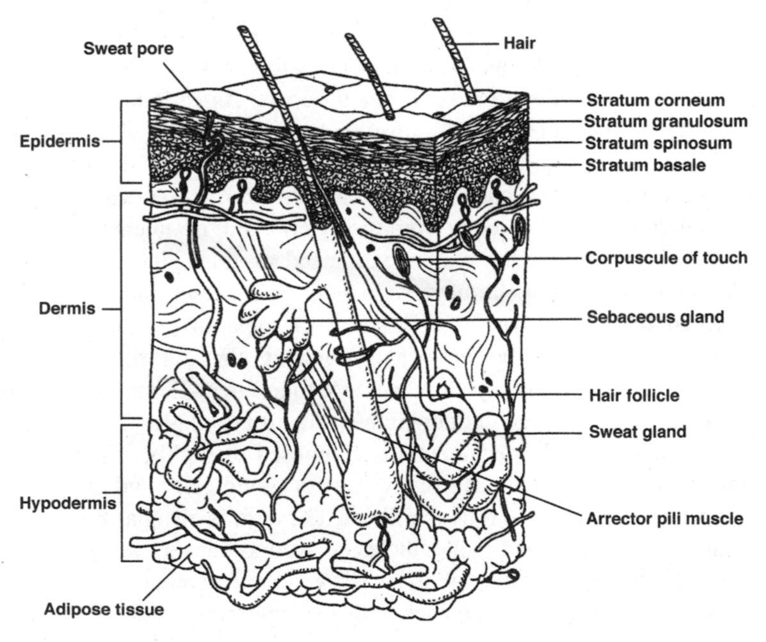


**Figure 5-1**. The skin

## You Need to Know 

The layers of the skin are, from superficial to deep, the:

- **Epidermis**
- **Dermis and Hypodermis**

The outer **epidermis** is composed of stratified squamous epithelium that is 30 to 50 cells thick. The layered cells are avascular; the outer cells are dead, keratinized, and cornified. The epidermis is stratified into five structural and functional layers, from superficial to deep, *stratum corneum*, *stratum lucidum*, *stratum granulosum*, *stratum spinosum*, and *stratum basale*. The stratum basale lies on the basement membrane of this epithelial tissue in close proximity to the underlying blood supply. Mitosis occurs primarily in the deep stratum basale and to a slight extent in the stratum spinosum. As the cells divide, only half of them remain in contact with the dermis. The other cells are pushed away from the underlying blood supply and cell death occurs. As cells move toward the surface, specialized cells, *keratinocytes*, fill with keratin (keratinization), a protein that toughens and waterproofs the skin, and all cells become flattened and scalelike (cornification). The dead cells of the epidermis buffer the body from the external environment.

Also found within the stratum basale and stratum spinosum are pigment forming cells, **melanocytes.** Melanin is a brown-black pigment produced by melanocytes. The amount of melanin produced varies among different ethnic groups. Other pigments that contribute to skin coloration are carotene, a yellow pigment found in epidermal cells, and hemoglobin, an oxygen binding pigment found in red blood cells.

The thick and deeper **dermis** is composed of highly vascularized connective tissue and consists of a variety of living cells, and numerous collagenous, elastic, and reticular fibers. The dermis also has numerous sweat and oil glands and hair follicles, as well as sensory receptors for heat, cold, touch, pressure, and pain. There are two layers of the dermis, the *papillary layer* is in contact with the epidermis and the deeper, thicker *reticular layer* is in contact with the hypodermis. Not considered a separate layer, the **hypodermis** (subcutaneous tissue) contains loose (areolar) connective tissue, adipose tissue, and blood and lymph vessels. Collagenous and elastic fibers reinforce the hypodermis. The hypodermis binds the dermis to underlying organs, stores lipids, insulates and cushions the body, and regulates temperature via autonomic vasoconstriction or vasodilation.

## Associated Structures of the Skin

*Hair, nails,* and three kinds of *exocrine glands* form from the epidermal skin layer. These structures develop as down-growths of germinal epidermal cells into the vascular dermis, where they receive sustenance and mechanical support.

The **hair follicle** is the germinal epithelial layer that has grown down into the dermis (Figure 5-2). Mitotic activity of the hair follicle accounts for growth of the hair. The **shaft of the hair** is the dead, visible, projecting portion; the **root of the hair** is the living portion within the hair follicle; and the **bulb of the hair** is the enlarged base of the root of the hair that receives nutrients and is surrounded by sensory receptors. The outer keratinized cuticle layer appears scaly under a dissecting microscope. Variation in the amount of melanin accounts for different hair colors. Each hair follicle has an associated **arrector pili** muscle (smooth muscle) that responds involuntarily to thermal or psychological stimuli, causing the hair to be pulled into a more vertical position. Hair on the scalp and eyebrows protects against the sunlight, hair in the nostrils and the eyelashes protect against airborne particles. A secondary function of hair is as a means of individual recognition and sexual attraction.

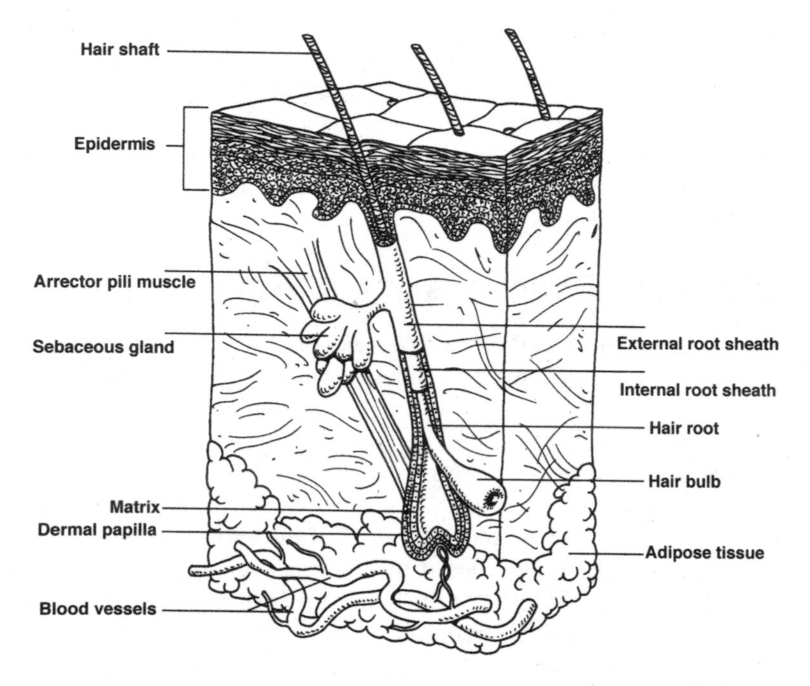


**Figure 5-2.** A hair within a hair follicle.

**Nails** are formed from the hardened, transparent stratum corneum of the epidermis. Nails serve to protect the digits and aid in grasping small objects. All lizards, birds, and mammals have some sort of hardened sheath (claw, talon, hoof, or nail) protecting their terminal phalanges.

The three types of **exocrine glands** are formed from the epidermal layer of skin.

- **Sebaceous glands:** Oil glands, secrete acidic sebum, lubricate and waterproof the skin.
- **Sudoriferous glands** are *sweat glands*. **Eccrine glands,** abundant on the forehead, back, palms, and soles, function in evaporative cooling. **Apocrine glands,** in the axillary and pubic regions, function at puberty as a sexual attractant.
- **Mammary glands** are specialized sudiferous glands within the breasts of females. Secrete milk under hormonal influence.

## Physiology of the Skin

**Table 5.1** Summary of the Physiology of the Skin

| Function | Site |
|---|---|
| Protects against:<br>Dehydration<br>Mechanical injury<br>Pathogens<br>Ultraviolet light | Epidermis |
| Blood loss | Epidermis and dermis |
| Synthesis of pigments and vitamin D | Epidermis and dermis |
| Temperature regulation via vasodilation, vasoconstriction, sweating and shivering | Dermis and hypodermis |
| Absorption of some $O_2$, $CO_2$, fat-soluble vitamins (A, D, E, and K); certain steroid hormones and some toxic substances | Epidermis, dermis, and hypodermis |
| Elimination of wastes: salts, water, and urea | Epidermis and dermis |
| Sensory reception for touch, temperature, pain, pressure, and stretch. | Epidermis, dermis, and hypodermis |

## Solved Problems

**True or False**

1. Skin is the largest tissue in the body, accounting for about 7 percent of body weight. (**True**)
2. Mitotic activity is characteristic of all layers of the epidermis except the dead stratum disjunction, which is constantly being shed. (**False**)
3. Mammary glands are modified sebaceous glands that are hormonally prepared to lactate in association with the birth of a baby. (**False**)

**Completion**

1. The dermis of the skin consists of an upper __________ layer and a deeper ___________ layer. (**papillary layer, reticular layer**)
2. The eipidermis of the skin consists of ______________ __________ epithelial tissue. (**stratified, squamous**)

# Chapter 6
# SKELETAL SYSTEM

IN THIS CHAPTER:

- ✔ *Structure and Function of Bone*
- ✔ *Bone Formation*
- ✔ *Bones of the Axial Skeleton*
- ✔ *Bones of the Appendicular Skeleton*
- ✔ *Articulations*
- ✔ *Solved Problems*

## Structure and Function of Bone

The skeletal system consists of bones, cartilage, and joints. Bones are composed of bone tissue, a connective tissue. The functions of the skeletal system fall into five categories.

1. **Support.** The skeleton forms a rigid framework to which are attached the softer tissues and organs of the body.
2. **Protection.** The skull, vertebral column, rib cage, and pelvic girdle enclose and protect vital organs; sites for blood cell production are protected with in the hollow centers of certain bones.
3. **Movement.** Bones act as levers when attached muscles contract, causing movement about joints.

4. **Hemopoiesis.** Red bone marrow of an adult produces white and red blood cells and platelets.
5. **Mineral and energy storage.** The matrix of bone is composed primarily of calcium and phosphorus; these minerals can be withdrawn in small amounts if needed elsewhere in the body. Lesser amounts of magnesium and sodium are also stored in bone tissue. Lipids stored in adipose cells of yellow bone marrow store energy.

## Categorization of Bones

| | |
|---|---|
| **Long bones:** | longer than wide, found in appendages. |
| **Short bones:** | more or less cubical, found in confined spaces. |
| **Flat bones:** | protection, bones of the skull, ribs. |
| **Irregular bones:** | odd shaped, vertebrae, certain skull bones. |

A long bone consists of a **diaphysis** (or shaft) in the center and an **epiphysis** on either end (Figure 6-1). Within the diaphysis is a **medullary cavity** that is lined with a thin layer of connective tissue, the **endosteum.** The medullary cavity contains fatty *yellow bone marrow*. The epiphyses consist of *spongy bone* surrounded by *compact bone*. *Red bone marrow* is found within the pores of the spongy bone. Separating the diaphysis and epiphysis is an **epiphyseal plate,** a region of mitotic activity responsible for linear bone growth (elongation); an **epiphyseal line** replaces the plate when bone growth is completed. A **periosteum** of dense regular connective tissue covers the bone and is the site of tendon-muscle attachment and diametric bone growth (widening).

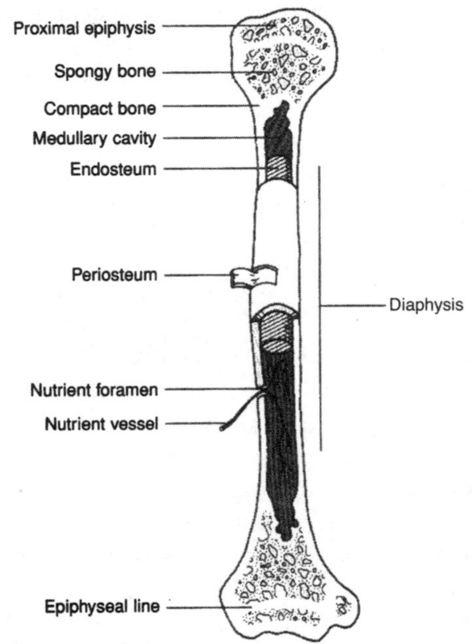


**Figure 6-1.** The structure of a long bone.

## Bone Formation

There are several different types of bone cells. **Osteogenic cells** are progenitor cells that give rise to all bone cells. **Osteoblasts** are the principal bone-building cells; they synthesize the collagenous fibers and bone matrix, and promote mineralization during ossification. The osteoblasts are then trapped in their own matrix and develop into **osteocytes** that maintain the bone tissue. **Osteoclasts** are bone-destroying cells that contain lysosomes and phagocytic vacuoles that demineralize bone tissue.

**Ossification** (bone formation) begins during the fourth week of prenatal development. Bones develop either through *endochondral* ossification—going first through a cartilaginous stage—or through *intramembranous* (dermal) ossification—forming directly as bone. **Endochondral ossification** of a long bone begins in a primary center in the

shaft of the cartilage model with hypertrophy of chondrocytes (cartilage cells) and calcification of the cartilage matrix. The cartilage model is then vascularized, osteogenic cells form a bony collar around the mode, and osteoblasts lay down bony matrix around the calcareous spicules. Ossification from primary centers occurs before birth; from secondary centers in the epiphyses, it occurs during the first 5 years. Most of the bones of the skeleton form through endochondral ossification.

The facial bones, most of the cranial bones, and the clavicle form via **intramembranous ossification.** During fetal development and infancy, the membranous bones of the top and sides of the cranium are separated by fibrous sutures. There are also six large membranous areas call **fontanels** that permit the skull to undergo changes in shape during birth. They also permit rapid growth of the brain during infancy. Ossification of the fontanels is complete by 20 to 24 month of age.

## Bones of the Axial Skeleton

The **axial skeleton** consists of the bones that form the axis of the body and that support and protect the organs of the head, neck, and trunk. These bone include the bones of the skull, vertebral column, rib cage, auditory ossicles and the hyoid bone.

The skull is composed of 8 *cranial bones* that articulate firmly with one another to enclose and protect the brain and associated sense organs, and 14 *facial bones* that form the foundation for the face and anchor the teeth. These bones are illustrated in Figure 6-2 and Figure 6-3.

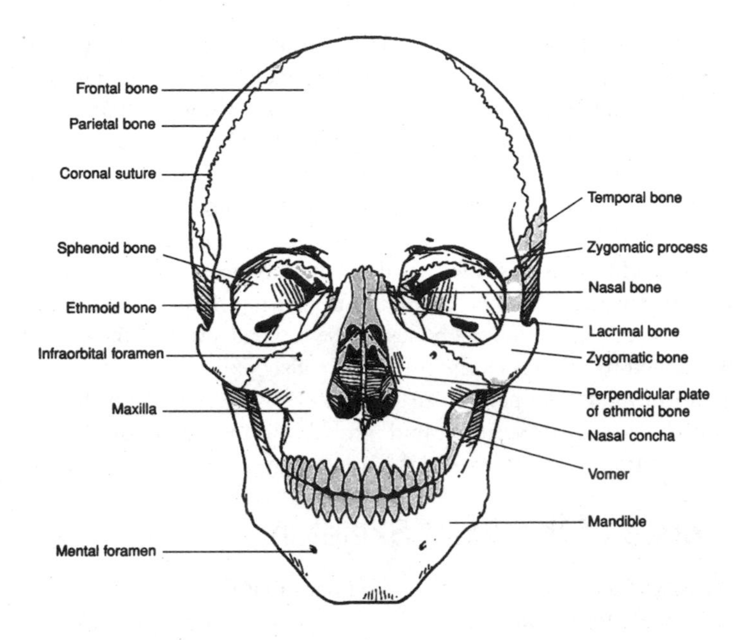


**Figure 6-2.** An anterior view of the skull.

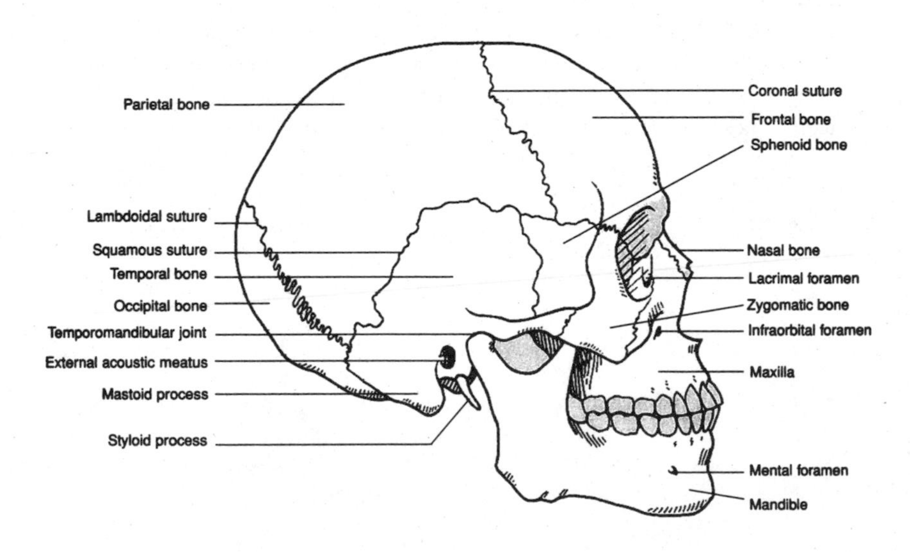


**Figure 6-3.** A lateral view of the skull.

The cranial bones consist of the **frontal bone** that forms the anterior roof of the cranium, two **temporal bones** that form the lower sides of the cranium, the **occipital bone** that forms the posterior and much of the inferior portion of the cranium, the **sphenoid bone** (which contains the sella turcica) forming the floor, and the **ethmoid bone** that contributes to the walls of the nasal cavity. The bones of the skull are joined by immovable joints called **sutures.** The **coronal suture** is at the junction of the frontal and two parietal bones, the **sagittal suture** between the two parietal bones, the **lambdoid suture** between the occipital and parietal bones, and the **squamous suture** is located between the temporal and parietal bones.

The facial bones are the **maxilla** (upper jaw), the **mandible** (lower jaw), the two **palatine bones** that contribute to the hard palate, the paired **zygomatic bones** (cheek bones), the paired **lacrimal bones** in the medial wall of the orbit, the two **nasal bones** forming the superior portion of the nose, the vomer and the paired **inferior nasal conchae,** which are located in the nasal cavity.

The **auditory** (ear) **ossicles** are contained within the petrous portion of the temporal bones. These three small bones, the **malleus** (hammer), **incus** (anvil), and **stapes** (stirrup) function to amplify and transmit sound from the outer ear to the inner ear. The **hyoid bone** is located in the anterior neck where it supports the tongue superiorly and the larynx (voice box) inferiorly.

The **vertebral column** (Figure 6-4) supports and permits movement of the head and trunk and provides a site for muscle attachment. The vertebrae also support and protect the spinal cord and permit the passage of spinal nerves.

## Remember

There are:

- **7 cervical vertebrae**
- **12 thoracic**
- **5 lumbar**
- **4 or 5 fused sacral**
- **4 or 5 fused coccygeal**

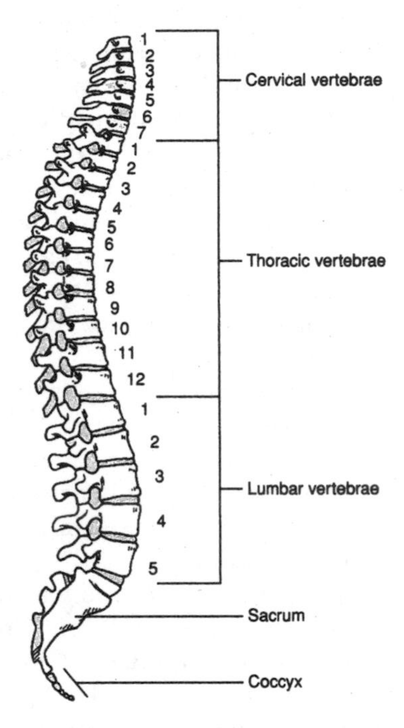


**Figure 6-4.** A lateral view of the vertebral column.

The vertebrae are separated by fibrocartilagenous **intervertebral discs** and are secured to one another by interlocking processes and ligaments. There is limited movement between vertebrae but extensive movements of the vertebral column as a unit. Between the vertebrae are openings called **intervertebral foramina** that permit passage of spinal nerves.

There are four curvatures in the adult vertebral column: the **cervical, thoracic,** and **lumbar curves** are designated by the type of vertebrae they include. The **pelvic curve** is formed by the sacrum and coccyx.

All vertebrae (Figure 6-5) have a **body,** a **neural arch** composed of two supporting **pedicles** and two arched **laminae,** a **vertebral foramen** that allows passage of the spinal cord, **spinous process,** paired **transverse processes,** paired **superior** and **inferior articular processes,** and an **intervertebral foramen** for passage of spinal nerves.

## Distinctive Features of Vertebrae

**Cervical vertebrae** → **transverse foramen**
**Thoracic vertebrae** → **facets for articulation with ribs**
**Lumbar vertebrae** → **broad, flat spineous process for muscle attachment**

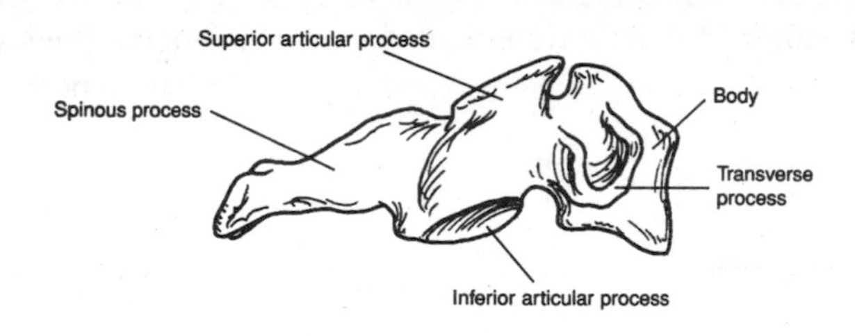


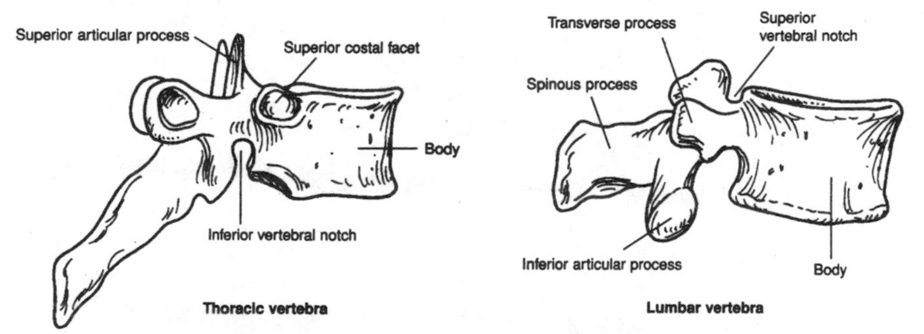


**Figure 6-5.** Examples of vertebrae from different vertebral regions.

The first cervical vertebra (C1) is called **atlas** and articulates with the occipital condyle of the skull, the second cervical vertebra (C2), **axis,** has a peglike **dens (odontoid) process** that provides a pivot for rotation with respect to atlas.

The remaining portion of the axial skeleton is the **rib cage** (Figure 6-6). The rib cage is composed of the **sternum, costal cartilages,** and the **ribs** attached to the *thoracic vertebrae*. The rib cage supports the pectoral girdle and the upper extremities, protects and supports the thoracic and upper abdominal viscera, provides an extensive surface area for muscle attachment, and plays a major role in respiration. The portions of the *sternum* (Figure 6-6) are the **manubrium,** the **body,** and the **xiphoid process.** Only the first 7 pairs of ribs are anchored to the sternum by individual costal cartilages, these are called the **true ribs** (vertebrosternal ribs). The remaining 5 pairs of ribs are called the **false ribs.** Ribs 8, 9, and 10 are attached to the costal cartilage of rib 7 (vertebrochondral ribs). Ribs 11 and 12 do not attach to the sternum at all and are referred to as the **floating ribs** (vertebral ribs). The first 10 ribs each have a **head** and **tubercle** for articulation with a vertebra (Figure 6-6). Ribs 11 and 12 have a head but no tubercle. All ribs have a **neck, angle,** and **shaft** (**body**).

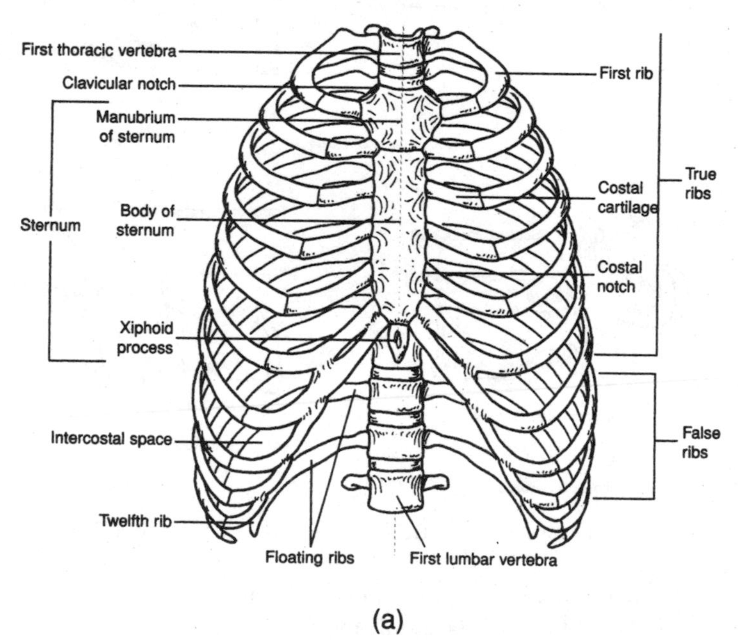


**Figure 6-6.** (a) The rib cage, anterior view.

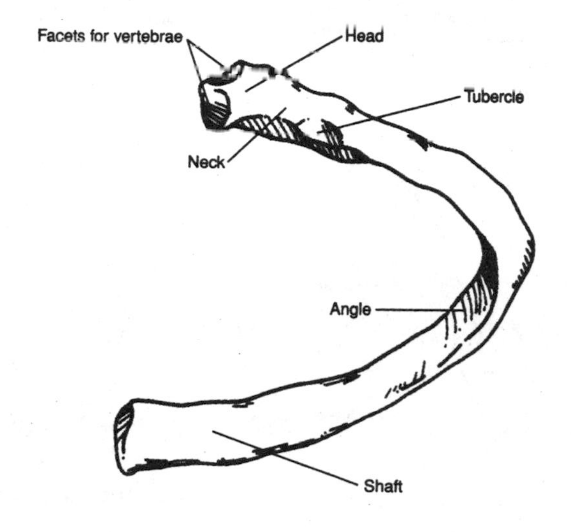


(b)

**Figure 6-6.** (b) A typical rib.

## Bones of the Appendicular Skeleton

The appendicular skeleton consists of bones of the *pectoral* and *pelvic girdles* and the bones of the *upper* and *lower extremities*. The girdles anchor the appendages to the axial skeleton.

The **pectoral girdle** consists of two scapulae and two clavicles that attach to the axial skeleton at the manubrium of the sternum. The pectoral girdle provides attachment for numerous muscles that move the brachium (arm) and antebrachium (forearm). The S-shaped **clavicle** binds the upper extremity to the axial skeleton and positions the shoulder joint away from the trunk for freedom of movement. Muscles of the trunk and neck attach to the clavicle. The **scapula** lies along the posterior thoracic wall and is attached to the axial skeleton via muscles. The features of the scapula are diagramed in Figure 6-7.

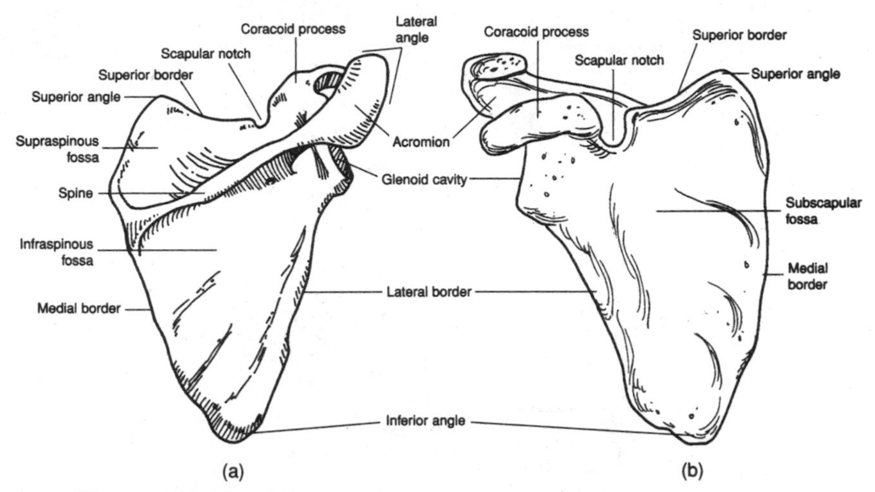


**Figure 6-7.** The right scapula (a) posterior view (b) anterior view

The upper extremity is divided into the *brachium*, that contains the *humerus*; the *antebrachium*, that contains the *radius* and *ulna* (Figures. 6-8 to 6-9); and the *manus* (hand), that contains 8 *carpal bones*, 5 *metacarpal bones*, and 14 *phalanges*. The features of the humerus (Figure 6-8), radius and ulna (Figure 6-9) are diagramed.

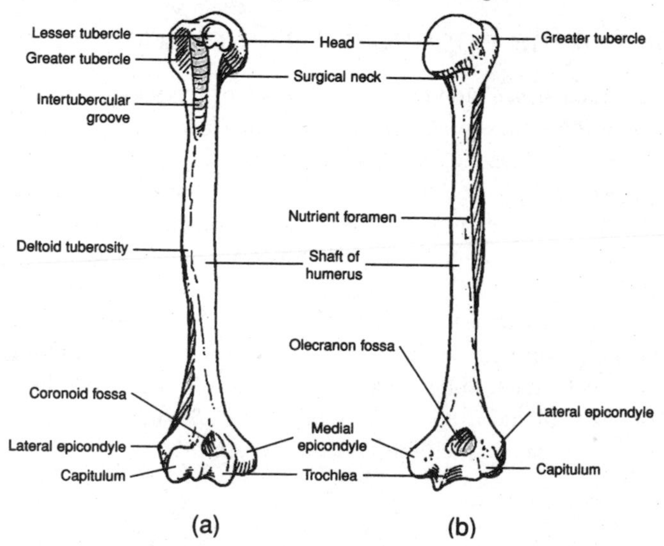


**Figure 6-8.** The right humerus (a) anterior view and (b) posterior view.

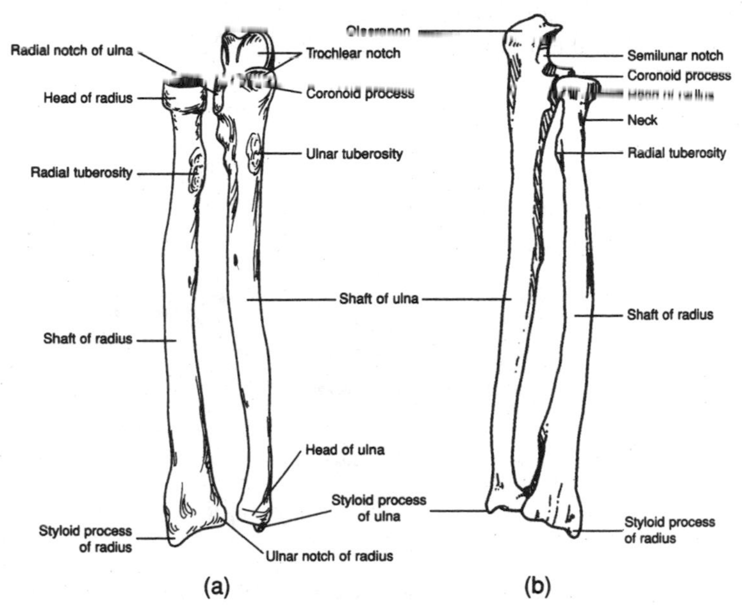


**Figure 6-9.** The right radius and ulna
(a) anterior view and (b) posterior view.

The **pelvic girdle,** or *pelvis*, is formed by the two **ossa coxae** united anteriorly by the **symphysis pubis**. It is attached posteriorly to the *sacrum* of the vertebral column at the *sacroiliac joints*. Each os coxae (hipbone) consists of an **ilium,** an **ischium,** and a **pubis.** In adults, these bones are firmly fused. The features of the os coxae are diagramed in Figure 6-10.

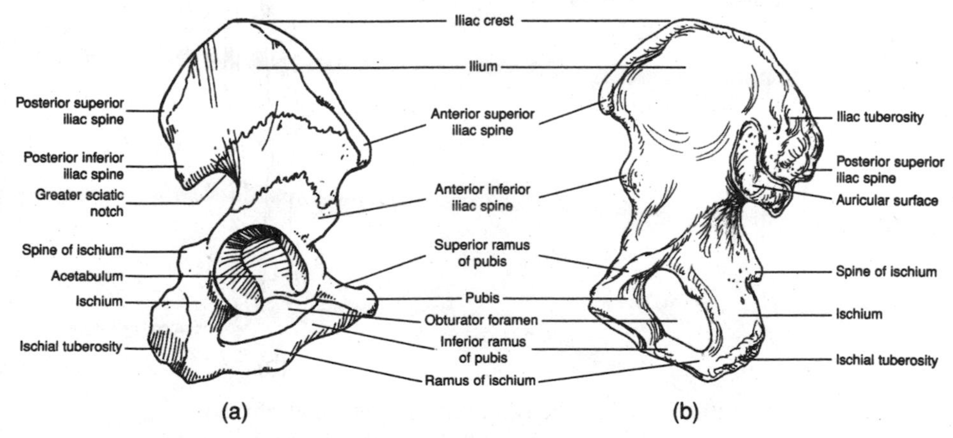


**Figure 6-10.** The right os coxae (a) lateral view and (b) medial view.

The female pelvis has a round or oval pelvic inlet, a wider pelvic outlet, and an obtuse pubic arch. These modifications are associated with child bearing.

The lower extremity is divided into the *thigh* and the *leg*. The **femur** is the only bone of the thigh. The **patella** (kneecap) is the sesamoid bone (formed in a tendon) of the anterior knee region. The **tibia** and **fibula** are the bone of the leg. The *distal end of the tibia* articulates with the *talus* in the *ankle*. The features of the femur, tibia and fibula are diagramed in Figures 6-11 and 6-12.

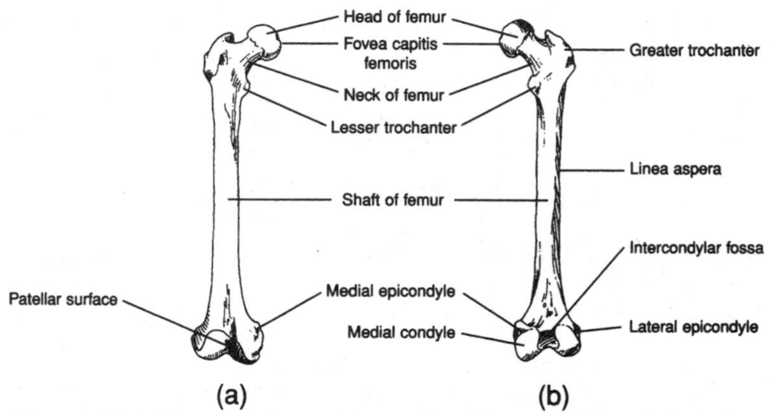


**Figure 6-11.** The right femur (a) anterior view and (b) posterior view.

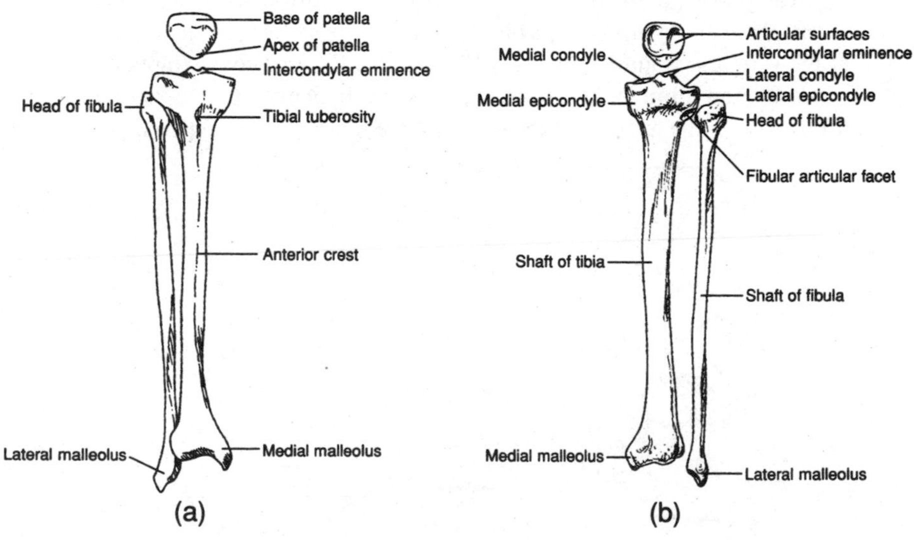


**Figure 6-12.** The right patella, tibia, and fibula (a) anterior view and (b) posterior view.

## Articulations

**Articulations,** or joints, may be classified according to structure or function. In the structural classification, a joint is *fibrous*, *cartilaginous*, or *synovial* (Table 6.1). The functional classification distinguishes *synarthroses* (immovable joints), *amphiarthroses* (slightly movable joints), and *diarthroses* (freely movable joints).

**Table 6.1** Structural Classification of Joints and Examples

| Classification | Structure | Examples |
|---|---|---|
| Fibrous joints | Articulating bones joined by fibrous connective tissue | • Sutures of the skull<br>• Tibia-fibula and radial-ulna joints<br>• Teeth in sockets |
| Cartilaginous joints | Articulating bones joined by fibrocartilage or hyaline cartilage | • Intervertebral joints<br>• Pubic symphysis<br>• Sacroiliac joint<br>• Epiphyseal plates |
| Synovial joints | Joint capsule containing synovial membrane and synovial fluid | • All freely moveable joints; most of the joints of the limbs |

All synovial joints are freely movable.

### Movements at Synovial Joints

**Flexion:** decreasing the angle between two bones;
**Extension:** increasing the angle between two bones;
**Abduction:** movement away from the midline of the body;
**Adduction:** movement toward the midline;
**Rotation:** movement of a bone around its own axis;
**Pronation:** rotation of the forearm that results in the palm directed backward;
**Supination:** the opposite rotation;
**Circumduction:** circular, conelike movement of a body segment.

## Solved Problems

**True or False**

____ 1. The proximal and distal ends of a long bone are referred to as diaphyses. (**False**)

____ 2. Cervical vertebrae are characterized by the presence of articular facets. (**False**)

____ 3. Most of the bones of the skeleton form through intramembranous ossification. (**False**)

____ 4. A person has seven pairs of true ribs and five pairs of false ribs, the last two pairs of which are designated as floating ribs. (**True**)

____ 5. Supination and pronation are specific kinds of circumductional movements. (**False**)

# Chapter 7
# Muscle Tissue and the Mode of Contraction

In This Chapter:

- ✔ *Microscopic Structure of Muscle*
- ✔ *Muscle Contraction*
- ✔ *Macroscopic Structure of Muscle*
- ✔ *Solved Problems*

There are three types of muscle tissue: smooth, cardiac, and skeletal. Each type has a different structure and function, and each occurs in a different location in the body.

Muscle functions include:

- **Motion.** Body movements such as walking, breathing, and speaking, as well as movements associated with digestion and the flow of fluids.
- **Heat production.**
- **Posture and body support.**

## Microscopic Structure of Muscle

Because muscle cells resemble tiny threads, they are called *muscle fibers*. Each skeletal muscle fiber is a multinucleated, striated cell containing a large number of rodlike **myofibrils** that extend the entire length of the cell. Each myofibril is composed of still smaller units, called **myofilaments** (or filaments). **Thin myofilaments** are composed primarily of the contractile protein **actin** and **thick myofilaments** contain primarily the contractile protein **myosin.**

### Structure of Myofilaments

Thick filament: Shaped like a golf club, each myosin protein has a long rod portion, and a globular head. The myosin head contains an actin binding site and a myosin ATPase binding site. The strands of the rod portion bind together with their globular head projecting outward to form the thick filaments that lie between the thin filaments (Figure 7-1).

**Thin filament:** The thin filament is composed of the proteins *actin*, *tropomyosin,* and *troponin*. Two long strands of **actin** molecules form the backbone of the thin myofilaments. Long, thin, threadlike **tropomyosin** proteins spiral around and cover the myosin binding sites on the actin helix. The **troponin** molecule fastens the ends of the tropomyosin molecule to the actin helix (Figure 7-2). In skeletal and cardiac muscle the thick and thin myofilaments overlap within the myofibril in a distinctive pattern called a **sarcomere.** The sarcomere is the structural and functional unit of the myofibril.

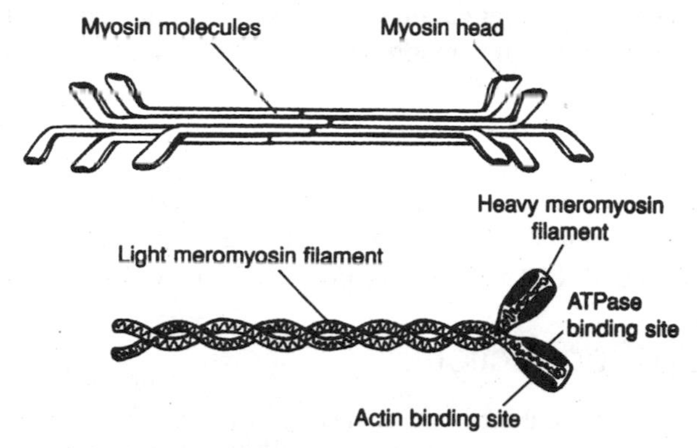


**Figure 7-1.** Structure of thick myofilaments.

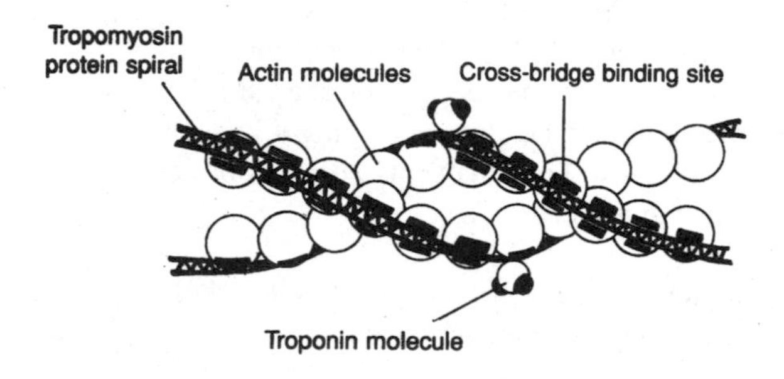


**Figure 7-2.** Structure of thin myofilaments.

The regular overlapping pattern of thin and thick filaments within the myofibrils is responsible for the cross-banding striations seen in skeletal and cardiac muscle. The dark bands contain the thick filaments and are called **A bands.** The lighter bands, the **I bands,** are regions where only thin filaments occur. The I bands are bisected by dark **Z lines** where the thin filaments of adjacent sarcomeres join.

## Structure of a Muscle Fiber (Cell)

The **sarcolemma** (cell membrane) of a muscle fiber encloses the **sarcoplasm** (cytoplasm). The sarcoplasm is permeated by a network of membranous channels, called the **sarcoplasmic** (endoplasmic) **reticulum,** that forms sleeves around the myofibrils. The longitudinal tubes of sarcoplasmic reticulum empty into expanded chambers called **terminal cisternae.** Calcium ions (Ca2+) are stored in the terminal cisternae and play an important role in regulating muscle contraction.

**Transverse tubules** (*T tubules*) are internal extensions of the sarcolemma that extend perpendicular to the sarcoplasmic reticulum. The T tubules pass between adjacent segments of terminal cisternae and penetrate deep into the interior of the muscle fiber to allow the action potential from the cell surface to be delivered into the center of the fiber.

## Muscle Contraction

In the sliding filament theory of contraction, a skeletal muscle fiber, together with all of its myofibrils, shortens by movement of the insertion toward the origin of the muscle. Shortening of the myofibrils is caused by shortening of the sarcomeres, which is accomplished by sliding of the myofilaments. The mechanism that produces the sliding of the thin (actin) myofilaments over the thick (myosin) myofilaments during contraction is outlined in the steps below.

1. Stimulation via the neurotransmitter acetylcholine across the neuromuscular junction initiates an action potential on the sarcolemma of the muscle fiber. This action potential spreads along the sarcolemma and is transmitted into the muscle fiber through the T tubules.
2. The T tubule potential causes the terminal cisternae of the sarcoplasmic reticulum to release calcium ions (Ca2+) in the immediate vicinity of each myofibril.
3. Ca2+ ions bind to and change the protein structure of the troponin molecules attached to the tropomyosin molecules on the thin filaments. This change causes the tropomyosin to move aside, exposing the actin binding sites.
4. Myosin cross bridges bind to actin. Upon binding, the cocked (energized) myosin head (HMM) undergoes a conformational change, causing the head to tilt. This pulls the actin filament over the myosin filament in an action called a *power stroke*.
5. After the power stroke, ATP binds the HMM, causing detachment of the cross bridge from the actin binding sites. The enzyme

ATPase within the HMM cleaves ATP to ADP + energy; the energy is used to recock the HMM. The HMM can then bind with another actin site (if still exposed due to the presence of Ca2+) and produce another power stroke.

6. Repeated power strokes successfully pull in the thin filaments. This sliding with a ratchet mechanism involves numerous actin binding sites and myosin cross bridges and constitutes a single muscle contraction.
7. Once the action potential ceases, the sarcoplasmic reticulum actively transports Ca2+ from the cytoplasm into the terminal cisternae. Without calcium ions, the troponin molecule resumes its original shape so that the tropomyosin is pulled back over the myosin binding sites of the thin filament. The thin filaments slide back to their noncontracted position, and the muscle is relaxed.

## Neuromuscular Junction

Stimulation from a motor neuron initiates the contraction of skeletal muscle. The space between the axon terminal of a motor neuron and the cell membrane of a muscle fiber is called the **neuromuscular** (myoneural) **junction** (Figure 7-3).

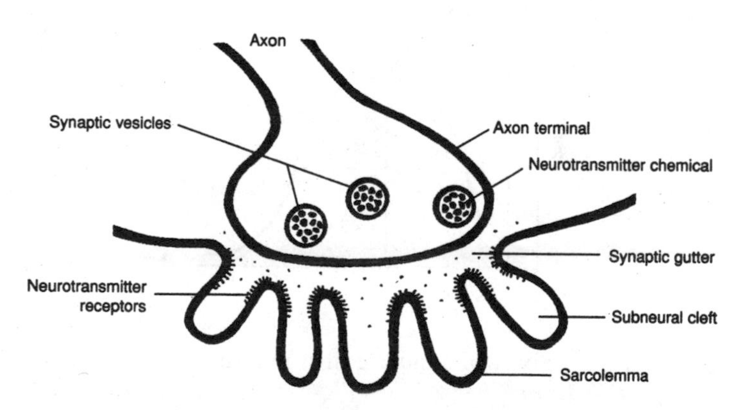


**Figure 7-3.** The neuromuscular junction.

The action potential travels along the motor neuron to the axon terminal, where it causes an influx of calcium ions. The calcium ions cause synaptic vesicles to release acetylcholine, which diffuses across the synaptic gutter and combines with specific receptors on the sarcolemma. An action potential radiates over the sarcolemma, initiating the sequence of events described above.

## Motor Unit

A **motor unit** consists of a single motor neuron together with the specific skeletal muscle fibers that it innervates. A large motor unit serves many muscle fibers, a small motor unit serves relatively few muscle fibers. Contraction of a skeletal muscle requires recruitment of motor units. Few motor units are recruited when fine, highly coordinated movements are being performed. Many motor units are recruited when strength is needed. Individual muscle fibers of a motor unit respond to an electrical stimulus in three phases (Figure 7-4):

1. the *latent period* between stimulation and the start of contraction;
2. the *contraction period*, or duration when work is being accomplished; and
3. the *relaxation period*, or recovery of the muscle fiber.

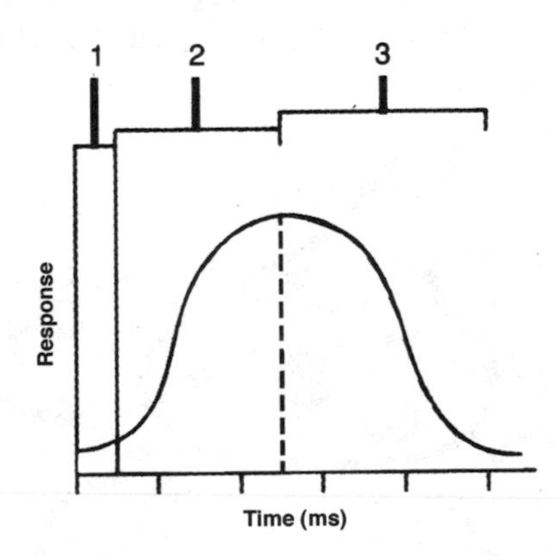


**Figure 7-4.** The activity of a muscle fiber in response to a stimulus.

**Important** ✓

## Types of Skeletal Muscle Fibers

**Fast-twitch fibers:** Large fibers with large amounts of glycogen; low amounts of $O_2$ carrying pigment myoglobin; anaerobic pathways to produce ATP; power and speed

**Slow-twitch fibers:** Small fibers with low glycogen content; high amount of myoglobin; aerobic pathways to produce ATP; resist fatigue, endurance.

**Intermediate fibers:** Intermediate in size and glycogen content; high amount of myoglobin; both aerobic and anaerobic pathways to produce ATP.

## Muscle Twitch, Summation, and Tetanus

A single action potential to the muscle fibers of a motor unit produces a muscle twitch, a rapid, unsustained contraction (Figure 7-5). If the impulses are applied to a muscle in rapid succession through several motor units, one twitch will not have completely ended before the next begins. Since the muscle is already in a partially contracted state when the second twitch begins, the degree of muscle shortening in the second contraction will be slightly greater than the shortening that occurs with a single twitch. The additional shortening due to the rapid succession of two or more action potentials is termed **summation**. At high stimulation frequencies, the overlapping twitches sum to one strong, steady contraction call **tetanus.**

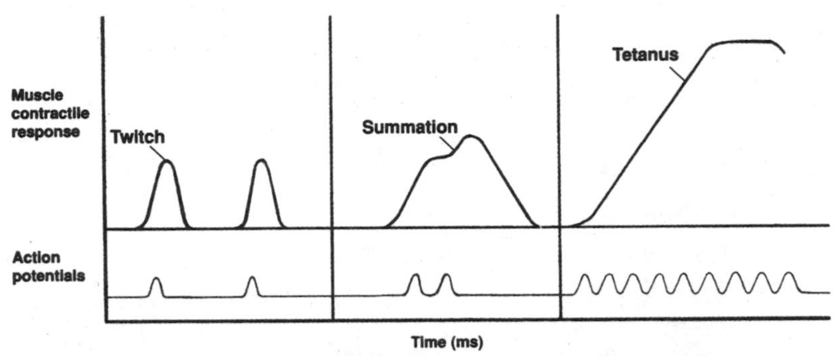


**Figure 7-5.** Patterns of muscle twitch, summation, and tetanus.

## Macroscopic Structure of Muscle

Skeletal muscle tissue, in association with connective tissue, is organized into muscle bundles. This muscle architecture determines the force and direction of the contracting muscle fibers. Muscle fibers may be organized with **parallel fibers, convergent fibers, pennate** (feather-shaped) **fibers,** or as a **sphincter** (circular) **muscle.**

Loose fibrous connective tissues bind muscles at various levels to unify the force of contraction. Surrounding each muscle fiber is a connective tissue called the **endomysium.** A group of individual muscle fibers is bound together by another connective tissue, the **perimysium,** to form a **fasciculus.** Many *fasciculi* make up an individual muscle. Each muscle is surrounded by a third connective tissue, the **epimysium.** These three connective tissues are continuous with the **tendon** that secures the muscle to bone.

Muscles are attached to the skeleton at two locations. The **origin** of a muscle is the more stationary attachment of the muscle; the **insertion** the more movable attachment. In the appendages, the origin is generally proximal in position and the insertion is distal in position.

## You Need to Know 

- the difference between muscle fibers, myofibrils, and myofilaments.
- the mechanism of muscle contraction and the importance of Ca2+ and ATP in this process.
- the organization and function of skeletal muscles: motor units, fiber types, types of contractions, and gross structure.

## Solved Problems

### Matching

____ 1. Z line (**a**)

____ 2. Sarcomere (**b**)

____ 3. A band (**f**)

____ 4. Sarcoplasmic reticulum (**c**)

____ 5. Troponin (**d**)

____ 6. Calcium (**e**)

____ 7. ATP-myosin complex (**g**)

(a) flat protein structure to which the thin filaments attach
(b) basic unit of a muscle fiber
(c) intramuscular saclike structures (tubules) derived from membranes
(d) structure that binds calcium
(e) “trigger” or regulator of contraction
(f) composed mainly of myosin molecules
(g) functions to release the energy in ATP

### True or False

____ 1. Actin is found only in the striated fibers of cardiac and skeletal muscle tissues. (**False**)

____ 2. Fasciculi are enclosed in a covering of perimysium. (**True**)

____ 3. Thin myofilaments are primarily composed of myosin proteins. (**False**)

# Chapter 8
# MUSCULAR SYSTEM

IN THIS CHAPTER:

- ✔ *Muscle Terminology*
- ✔ *Muscles of the Axial Skeleton*
- ✔ *Muscles of the Appendicular Skeleton*
- ✔ *Solved Problems*

## Muscle Terminology

Muscles are named according to various features of the muscle.

Muscle Nomenclature

| Feature | Example |
|---|---|
| Shape | Deltoid (like a triangle) |
| Location | Pectoralis (chest region) |
| Attachment(s) | Sternocleidomastoid |
| Orientation | Rectus (straplike) |
| Relative position | Lateralis; medialis |
| Function | Abductor, flexor |

The function (or action) of a muscle is described using the terms below.

| **Action** | **Definition** |
|---|---|
| Flexion | Decreases the joint angle |
| Extension | Increases a joint angle |
| Abduction | Movement away from the midline |
| Adduction | Movement toward the midline |
| Elevation | Raises a body structure |
| Depression | Lowers a body structure |
| Rotation | Turns a bone around its longitudinal axis |
| Supination | Rotates the forearm so palm faces anteriorly |
| Pronation | Rotates the forearm so palm faces posteriorly |
| Inversion | Turns the sole inward |
| Eversion | Turns the sole outward |

## Muscles of the Axial Skeleton

The muscles of the axial skeleton include those used in facial expression (not discussed here), mastication, neck movement, and respiration; those that act on the abdominal wall and those that move the vertebral column.

The **muscles of mastication** that close the jaw are the:

| | | |
|---|---|---|
| Temporalis | O: temporal fossa | I: coronoid process of mandible |
| Masseter | O: zygomatic arch | I: lateral ramus of mandible |
| Medial pterygoid | O: sphenoid bone | I: medial ramus of mandible |
| Lateral pterygoid | O: sphenoid bone | I: anterior side of mandibular condyle |

The **muscles of neck movement** include:

| | | |
|---|---|---|
| Sternocleidomastoid | O: sternum and clavicle | I: mastoid process |
| Digastric | O: mastoid process; inferior border of the mandible | I: hyoid bone |

The **muscles of the abdominal wall** all function to compress the abdomen, some act in lateral rotation and flexing the vertebral column.

| | | |
|---|---|---|
| External oblique | O: lower eight ribs | I: iliac crest and linea alba |
| Internal oblique | O: iliac crest, inguinal ligament, lumbar fascia | I: linea alba; costal cartilages of lower ribs |
| Transversus abdominis | O: iliac crest, inguinal ligament, lower ribs | I: xiphoid process, linea alba, pubis |
| Rectus abdominis | O: pubic crest; pubic symphysis | I: xiphoid process; costal cartilages ribs 5 – 7. |

The **muscles of the vertebral column** include a group of muscles collectively called the **erector spinae** muscles. These muscles are longitudinally directed muscles that function to extend the vertebral column.

## Muscles of the Appendicular Skeleton

The muscles of the appendicular skeleton include those of the pectoral girdle, brachium (arm), antebrachium (forearm), manus (hand), thigh, leg, and pes (foot).

The **muscles of the pectoral girdle** attach the pectoral girdle to the axial skeleton and are involved in movements of the scapula.

| | | |
|---|---|---|
| Serratus anterior | O: upper 8 or 9 ribs | I: anterior medial border of the scapula |
| Pectoralis minor | O: sternal ends of ribs 3 – 5 | I: coracoid process of scapula |
| Trapezius | O: Occipital bone and spines of cervical and thoracic vertebrae | I: clavicle, acromion and spine of scapula |
| Levator scapulae | O: 1$^{st}$ to 4$^{th}$ cervical vertebrae | I: superior border of scapula |
| Rhomboideus major | O: spines of thoracic vertebrae 2 – 5 | I: medial border of scapula |
| Rhomboideus minor | O: 7$^{th}$ cervical and 1$^{st}$ thoracic vertebrae | I: medial border of scapula |

The **muscles that move the humerus** at the shoulder joint function to flex, extend, rotate, abduct or adduct the humerus. The function of each muscle can be determined by the origin, insertion of each muscle.

| | | |
|---|---|---|
| Pectoralis major | O: clavicle, sternum, ribs 2 – 6 | I: greater tubercle of humerus |
| Latissimus dorsi | O: spines of sacral, lumbar, and lower thoracic vertebrae | I: intertubercular groove of humerus |
| Deltoid | O: clavicle, acromion, and spine of scapula | I: deltoid tuberosity |
| Supraspinatus | O: supraspinous fossa | I: greater tubercle of humerus |
| Infraspinatus | O: infraspinous fossa | I: greater tubercle of humerus |
| Teres major | O: inferior angle and lateral border of scapula | I: intertubercular groove of humerus |
| Teres minor | O: lateral border of scapula | I: greater tubercle of humerus |
| Subscapularis | O: Subscapular fossa | I: lesser tubercle of humerus |
| Coraco-brachialis | O: Coracoid process | I: shaft of humerus |

The ventral **muscles that act on the antebrachium** (forearm) function to flex the elbow; the dorsal muscles function to extend the elbow.

| | | |
|---|---|---|
| Biceps brachii | O: coracoid process and tuberosity above glenoid fossa of scapula | I: radial tuberosity |
| Brachialis | O: anterior shaft of humerus | I: coronoid process of ulna |
| Brachioradialis | O: lateral supracondylar ridge of humerus | I: distal radius |
| Triceps brachii | O: tuberosity below glenoid fossa and shaft of the humerus | I: olecranon of ulna |
| Anconeus | O: lateral epicondyle of humerus | I: olecranon of ulna |

There are numerous **muscles that act on the wrist, hand, and fingers.** They can be divided into three general groups:

1. **Pronators and Supinators,** function to pronate or supinate the forearm;
2. **Flexors,** flex the wrist and digits; and
3. **Extensors,** extend the wrist and digits.

The pronator and all the flexors arise from the medial epicondyle of the humerus; the supinator and extensors all originate from the lateral epicondyle of the humerus.

The **anterior muscles that move the thigh** at the hip all function to flex and laterally rotate the hip. The **posterior muscles** extend, abduct, and some medially rotate the hip.

| | | |
|---|---|---|
| Iliacus | O: iliac fossa | I: lesser trochanter of femur |
| Psoas major | O: transverse process of lumbar vertebrae | I: lesser trochanter of femur |
| Gluteus maximus | O: iliac crest, sacrum, coccyx, aponeurosis of lumbar region | I: gluteal tuberosity and iliotibial tract |
| Gluteus medius | O: lateral surface of ilium | I: greater trochanter of femur |
| Gluteus minimus | O: lateral surface of ilium | I: greater trochanter of femur |
| Tensor fasciae latae | O: anterior border of ilium and iliac crest | I: iliotibial tract |

The **medial muscles that move the thigh** at the hip all function to adduct the thigh.

| | | |
|---|---|---|
| Gracilis | O: symphysis pubis | I: proximomedial surface of tibia |
| Pectineus | O: pectineal line of pubis | I: distal to lesser trochanter of femur |
| Adductor longus | O: pubis | I: linea aspera |
| Adductor brevis | O: pubis | I: linea aspera |
| Adductor magnus | O: inferior rami of pubis and ischium | I: linea aspera and medial epicondyle of femur |

The **muscles of the thigh that move the leg** are divided into an **anterior group** that primarily function to extend the leg at the knee (the exception is sartorius, which flexes the leg and thigh), and a **posterior group** (the hamstrings) that functions to extend the thigh at the hip and flex the leg at the knee.

| | | |
|---|---|---|
| Sartorius | O: anterior superior iliac spine | I: medial surface of tibia |
| Rectus femoris | O: anterior inferior iliac spine | I: patella by common tendon |
| Vastus lateralis | O: greater trochanter and linea aspera | I: patella by common tendon |
| Vastus medialis | O: medial surface and linea aspera | I: patella by common tendon |
| Vastus intermedius | O: anterior and lateral surfaces of femur | I: patella by common tendon |
| Biceps femoris | O: ischial tuberosity; linea aspera | I: head of fibula and lateral epicondyle tibia |
| Semitendinosus | O: ischial tuberosity | I: proximal portion of shaft of tibia |
| Semimembranosus | O: ischial tuberosity | I: medial epicondyle of tibia |

The **muscles of the leg that move the ankle, foot, and toes** are separated into (1) an anterior group (including tibialis anterior) that dorsiflexes the foot and extends the digits, (2) a lateral group (the peroneal muscles) that aid in dorsiflexion and eversion, and (3) a posterior group (including gastrocnemius and soleus) that plantar flexes the foot and flexes the toes. Only a few are described here.

| | | |
|---|---|---|
| Tibialis anterior | O: lateral condyle of tibia | I: 1st metatarsal and tarsal |
| Gastrocneumius | O: lateral and medial epicondyle of femur | I: posterior surface of calcaneus |
| Soleus | O: posterior aspect of fibula and tibia | I: calcaneus |

For each muscle you should know the:

- **Origin**
- **Insertion**
- **Action**

## Solved Problems

1. A flexor muscle of the shoulder joint is (*a*) the supraspinatus, (*b*) the trapezius, (*c*) the pectoralis major, (*d*) the teres major. (**c**)
2. Which of the following muscles does not attach to the humerus? (*a*) teres major, (*b*) supraspinatus, (*c*) biceps brachii, (*d*) brachialis, (*e*) pectoralis major. (**c**)
3. Of the four quadriceps femoris muscles, which contracts over the hip and knee joints? (*a*) rectus femoris, (*b*) vastus medialis, (*c*) vastus intermedius, (*d*) vastus lateralis. (**a**)
4. Which of the following muscles does *not* attach to the rib cage? (*a*) serratus anterior, (*b*) rectus abdominis, (*c*) pectoralis major, (*d*) latissiumus dorsi. (**d**)
5. Which of the following is *not* used as a means of naming muscles? (*a*) location, (*b*) action, (*c*) shape, (*d*) attachment, (*e*) strength of contraction. (**e**)
6. A muscle of mastication is (*a*) the buccinator, (*b*) the temporalis, (*c*) the mentalis, (*d*) the zygomaticus, (*e*) the orbicularis oris. (**b**)

# Chapter 9
# NERVOUS TISSUE

IN THIS CHAPTER:

- ✔ *The Nervous System*
- ✔ *Neurons and Neuroglia*
- ✔ *Physiology of Nerve Conduction*
- ✔ *Synapse and Synaptic Transmission*
- ✔ *Solved Problems*

## The Nervous System

On the basis of structure, the nervous system is divided into the **central nervous system** (**CNS**) and the **peripheral nervous system** (**PNS**). The CNS is composed of the *brain* and the *spinal cord*. The PNS is composed of *cranial nerves* from the brain and *spinal nerves* from the spinal cord. In addition, *ganglia*, clusters of cell bodies of neurons, and *plexuses*, networks of nerves, are found within the PNS.

The **autonomic nervous system** (**ANS**) is a functional division of the nervous system. Structures within the brain are ANS control centers, and specific nerves are the pathways for conduction of autonomic nerve impulses. The ANS functions automatically to speed up or slow down body activities.

### Functions of the Nervous System

- Respond to internal and external stimuli.
- Transmit nerve impulses to and away from CNS.
- Interpret nerve impulses at the cerebral cortex.
- Assimilate experiences in memory and learning.
- Initiate glandular secretion and muscle contraction.
- Program instinctual behavior.

## Neurons and Neuroglia

A **neuron** is a nerve cell found in both the CNS and the PNS. Although neurons vary considerably in size and shape, they are generally composed of a **cell body, dendrites,** and an **axon** (Figure 9-1).

At the ends of the branched axon are slight enlargements, **axon terminals,** that contain **synaptic vesicles** that produce and secrete *neurotransmitter chemicals* in the *synapses.*

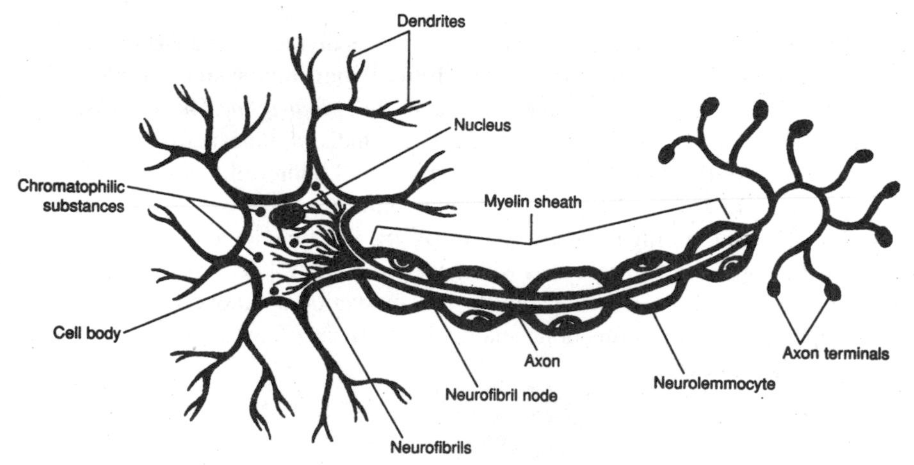


**Figure 9-1.** The structure of a neuron.

## You Need to Know

### Types of Neurons

- **Sensory neurons:** transmit impulses <u>to</u> the CNS.
  **somatic sensory:** carry impulses from receptors in the skin, bones, muscles and joints.
  **visceral sensory:** carry impulses from the visceral organs.
- **Motor neurons:** transmit impulses <u>away</u> form the CNS.
  **somatic motor:** innervate skeletal muscles.
  **visceral motor (autonomic motor):** innervate cardiac muscle, smooth muscle, and glands.
- **Association neurons (interneurons):** conduct impulses from sensory to motor neurons.

**Myelin** is an insulating sheath of a fatlike lipid that wraps around the axon of many neurons. This sheath is produced by specific **neuroglia** cells. In the PNS, there are small gaps between segments of the sheath. The myelin sheath insulates nerve fibers and speeds up transmission of an impulse along the axon.

**Neuroglia** are specialized cells of the nervous support neurons. There are six different types of neuroglia, all mitotically divide, and are about five times more abundant than neurons.

## Physiology of Nerve Conduction

In a non-conducting ("resting") neuron, a voltage, or **resting potential,** exists across the cell membrane. This resting potential is due to an imbalance of charged particles (ions) between the extracellular and the intracellular fluids. The mechanisms responsible for the membrane having a net positive charge on its outer surface and a net negative charge on its inner surface (Figure 9-2) are as follows:

1. A *sodium-potassium pump* actively transports sodium ions (Na+) to the outside and potassium ions (K+) to the inside, with three Na+ moved out for every two K+ moved in.

2. The cell membrane is more permeable to K+ than to Na+, so that the K+ , which is more concentrated inside the cell, diffuses outward faster than the Na+, which is more concentrated outside the cell, diffuses inward. Na+ and K+ move through the membrane using different channels.
3. The cell membrane is essentially impermeable to the large (negatively charged) anions that are present inside the neuron, therefore fewer negatively charged particles move out than positively charged particles.

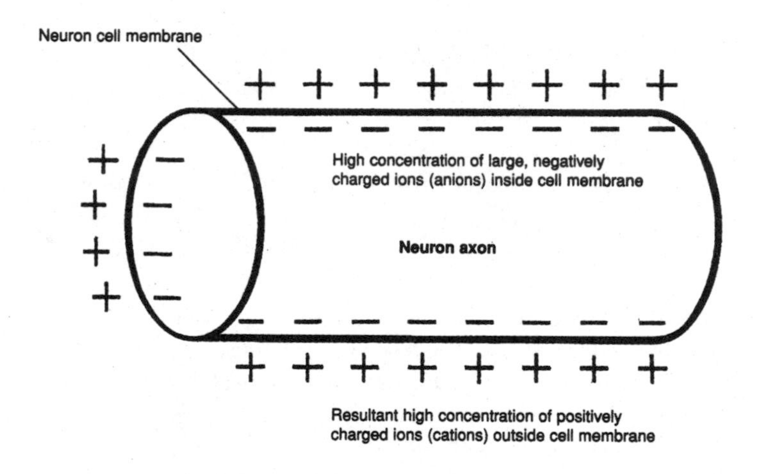


**Figure 9-2.** A segment of a neuron showing the location of charges.

Nerve impulses carry information from one point of the body to another by progression along the neuron membrane of an abrupt change in the resting potential. This "traveling disturbance," called an **action potential,** is described below.

1. A stimulus (chemical-electrical-mechanical) is sufficient to alter the resting membrane potential of a region of the membrane.
2. The membrane's permeability to sodium ions (Na+) increases at the point of stimulation.
3. Na+ rapidly moves into the cell through the membrane; the membrane becomes locally depolarized (membrane potential = 0).
4. Na+ continues to move inward; the inside of the membrane becomes positively charged relative to the outside (reverse polarization).

5. Reverse polarization at the original site of stimulation results in a local current that acts as a stimulus to the adjacent region of the membrane.
6. At the point originally stimulated, the membrane's permeability to sodium decreases, and its permeability to K+ increases.
7. K+ rapidly moves outward, again making the outside of the membrane positive in relation to the inside (repolarization).
8. N+ and K+ pumps transport Na+ back out of, and K+ back into the cell. The cycle repeats itself, traveling in this manner along the neuron membrane.

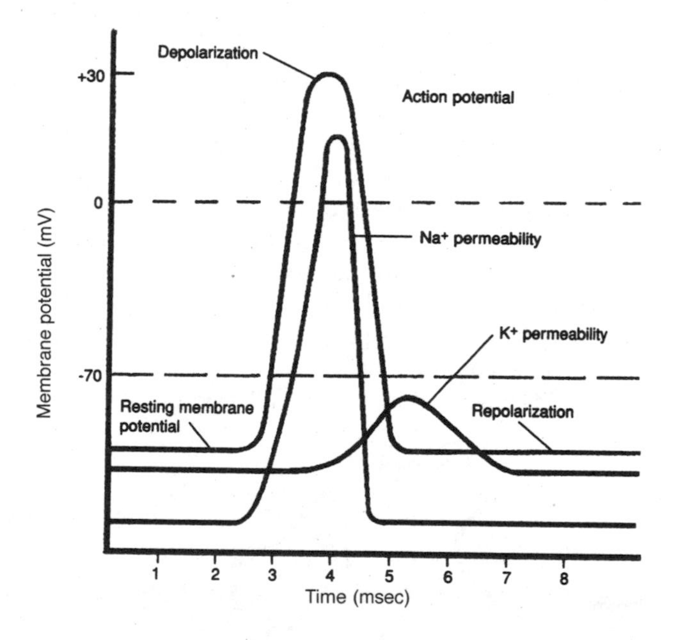


**Figure 9-3.** An action potential.

An action potential will be produced in response to a *threshold stimulus*. The resting membrane potential is about –70mV. If a stimulus raises the membrane potential –55 mV, a *threshold potential* has been reached, complete depolarization and repolarization occur, and an action potential is generated. (See Figure 9-3.)

> **All-or-None Law**
>
> A threshold stimulus evokes a maximal response and a sub-threshold stimulus evokes no response.

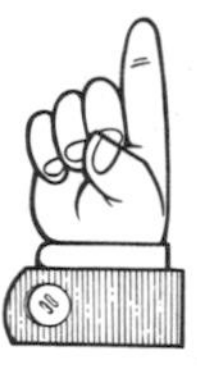

## Synapse and Synaptic Transmission

A **synapse** is the specialized junction through which impulses pass from one neuron to another (*synaptic transmission*), via chemical messengers (**neurotransmitters**). Refer to Figure 9-4 and steps below.

1. An action potential reaches the axon terminal.
2. An influx of Ca2+ causes synapatic vesicles containing neurotransmitter to fuse with the presynaptic membrane.
3. Neurotransmitter is released by exocytosis into the synaptic cleft.
4. The neurotransmitter diffuses across the cleft to the postsynaptic membrane and bind to specific receptors located there.
5. The permeability of postsynaptic membrane is altered, initiating an impulse on the second neuron.
6. The neurotransmitter is removed from the synapse.

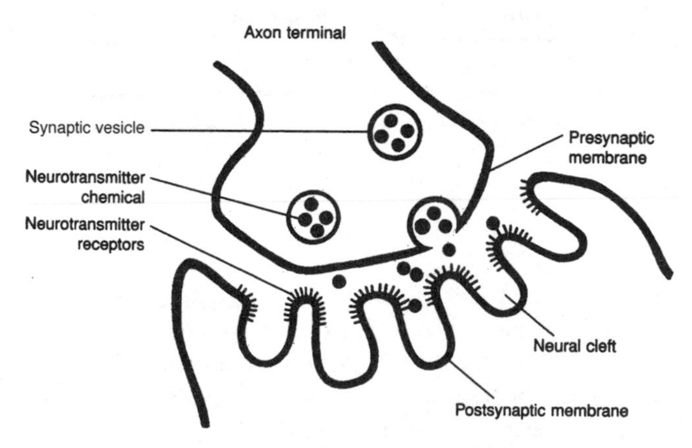


**Figure 9-4.** Synaptic transmission.

Neurotransmitters may be excitatory, causing the postsynaptic neuron to become active by producing an *excitatory postsynaptic potential* (EPSP), or may be inhibitory, preventing the postsynaptic neuron from becoming active by producing an *inhibitory postsynaptic potential* (ISPS).

## Solved Problems

### True or False

____ 1. There are basically only two different cell types in the nervous system. (**False**)

____ 2. A polarized nerve fiber has an abundance of sodium ions on the outside of the axon membrane. (**True**)

____ 3. All synapses are inhibitory. (**False**)

____ 4. The myelin sheath surrounds the dendrites. (**False**)

____ 5. Motor neurons convey information from receptors in the periphery to the CNS. (**False**)

____ 6. The sodium pumps operate by diffusion, and thus requires no ATP for its operation. (**False**)

### Completion

1. Within the peripheral nervous system, myelin is formed by the ______________. (**neurolemmocytes or Schwann cells**)
2. A junction between two neurons, where the electrical activity of the first influences the excitability of the second is called a ______________. (**synapse**)
3. When an action potential depolarizes the synaptic knob, small quantities of transmitter substance are released into the ______________ ______________. (**synaptic cleft**)

# Chapter 10
# Central Nervous System

In This Chapter:

- ✔ *Brain*
- ✔ *Meninges*
- ✔ *Blood-Brain Barrier*
- ✔ *Neurotransmitters*
- ✔ *Spinal Cord*
- ✔ *Solved Problems*

The **central nervous system (CNS)** consists of the *brain* and *spinal cord*. The functions of the CNS include body orientation and coordination, assimilation of experiences (learning), and programming of instinctual behavior. The CNS contains gray and white matter. The **gray matter** consists of either nerve cell bodies and dendrites, or of unmyelinated axons and neuroglia. It forms the outer convoluted cerebral cortex and cerebellar cortex in the brain and forms the inner portion of the spinal cord. The **white matter** consists of aggregations of myelinated axons and forms nerve **tracts,** within the CNS.

## Brain

There are five regions of the brain, some with multiple structures:

| Brain region | Structures |
|---|---|
| Telencephalon | Cerebrum |
| Diencephalon | Thalamus, Hypothalamus, and Pituitary gland |
| Mesencephalon | Superior colliculus, Inferior colliculus, and Cerebral peduncles |
| Metencepahlon | Cerebellum and Pons |
| Myelencephalon | Medulla oblongata |

**Telencephalon—Cerebrum**

The cerebrum consists of five paired lobes within two convoluted **cerebral hemispheres.** The hemispheres are connected by the **corpus callosum.** The cerebrum is responsible for higher functions, including perception of sensory impulses, instigation of voluntary movement, memory, thought, and reasoning. The outer convoluted surface, the **cerebral cortex,** is composed of gray matter. Elevated folds of the convolutions are the **gyri** (*gyrus*, singular) and the depressed grooves are the **sulci** (*sulcus*, singular). The convolutions greatly increase the surface area of the gray matter. Beneath the cerebral cortex is the thick white matter, the **cerebral medulla.**

## You Need to Know ✓

### 5 Cerebral Lobes and Their Functions:

| | |
|---|---|
| **Frontal lobe** | Voluntary control of skeletal muscles; personality; intellectual process; verbal communication. |
| **Parietal lobe** | Cutaneous and muscular sensations; understanding and utterance of speech. |
| **Temporal lobe** | Interpretation of auditory sensations; auditory and visual memory. |
| **Occipital lobe** | Integration of movements in focusing the eye; correlation of visual images with previous experiences; conscious seeing. |
| **Insular** | Memory; integration of other cerebral activities. |

### Diencephalon

The diencephalon, a major autonomic region of the forebrain, is almost completely surrounded by the cerebral hemispheres. It contains the:

- **Thalamus.** The thalamus is a paired organ immediately below the lateral ventricle. It is a relay center for all sensory impulses, except smell, to the cerebral cortex.
- **Hypothalamus.** The hypothalamus consists of several nuclei interconnected to other parts of the brain. Most of its functions relate to regulation of visceral activities including: cardiovascular regulation, body-temperature regulation, water and electrolyte balance, gastrointestinal activity and hunger, sleeping and wakefulness, sexual response, emotions, and control of endocrine functions through stimulation of the anterior pituitary.
- **Epithalamus.** The **pineal gland** extends from the epithalamus. It secretes the hormone *melatonin,* which may play a role in the onset of puberty.
- **Pituitary Gland.** The pituitary is divided into the anterior pituitary, the **adenohypophysis,** and the posterior pituitary, the **neurohypophysis**. This gland has endocrine functions.

### Mesencephalon

The mesencephalon, or midbrain, is a short section of the brain stem between the diencephalon and the pons. It contains the **superior colliculi,** concerned with visual reflexes, the **inferior collicui,** responsible for auditory reflexes, and the **cerebral peduncles,** which contain sensory and motor fibers and are involved with coordinating reflexes.

### Metencephalon

The metencephalon contains the:

- **Pons.** The pons consists of fiber tracts that relay impulses from one region of the brain to another. Many cranial nerves originate here. Also the *apneusitic* and *pneumotaxic* centers involved with regulating respiratory rate are located in the pons.
- **Cerebellum.** The cerebellum consists of two hemispheres and is responsible for involuntary coordination of skeletal-muscle contractions in response proprioreceptors within muscles, tendons, joints, and sensory organs.

**Mylencephalon—Medulla oblongata**

The medulla oblongata connects to the spinal cord and makes up much of the brain stem. It is composed primarily of white matter tracts that communicate between the spinal cord and the brain. Three areas that control autonomic functions are: the **cardiac center,** sending inhibitory and accelerator fibers to the heart; the **vasomotor center,** which causes the smooth muscle of arterioles to contract; and the **respiratory center,** which controls the rate and depth of breathing.

**Ventricles of the brain**

The ventricles of the brain consist of a series of cavities that are connected to one another and to the central canal of the spinal cord.

| | |
|---|---|
| **Lateral ventricles** | Located in each cerebral hemisphere |
| **Third ventricle** | Located in the diencephalon |
| **Fourth ventricle** | Located in the brain stem |

## Meninges

Three connective tissue membranes cover the entire CNS.

The three meninges are (from outermost to innermost): **Dura mater, Arachoid, Pia mater.**

The dura mater forms a tough tubular sheath around the spinal cord. The **epidural space** is a vascular area between the sheath and the vertebral canal. It is the location where epidural anesthesia is injected. The **subarachnoid space** is located between the arachnoid and the pia mater. It contains *cerebrospinal fluid.* **Cerebrospinal fluid (CSF)** is a clear, lymph-like fluid produced continually by active transport of substances from blood plasma by specialized capillaries in the roof of the ventricles, the *choroid plexuses.* CSF forms a protective cushion around and within the CNS; it also buoys the brain. CSF circulates around the CNS through the ventricles of the brain, the central canal of the spinal cord, and the subarachnoid space.

## Blood-Brain Barrier

The **blood-brain barrier (BBB)** is a structural arrangement of the capillaries that surround connective tissue and the vascular processes of *astrocytes* (a type of neuroglia in the CNS) that cling to the capillaries. The BBB selectively determines which substances can move from the blood plasma to the extracellular fluid of the brain. Fat-soluble compounds readily pass through the BBB, as do $H_2O$, $O_2$, $CO_2$, and glucose. Certain chemicals such as alcohol, nicotine, and anesthetics also readily pass through. Inorganic ions pass more slowly and other substances, such as macroproteins, lipids, certain toxins, and most antibiotics are restricted.

## Neurotransmitters

There are over 200 neurotransmitters synthesized and secreted by neurons within the brain. The most important are listed below.

| **Neurotransmitter** | **Function** |
|---|---|
| Acetylcholine | Transmits impulses across synapses |
| Epinephrin, norepinephrin | Arouse the brain and maintain alertness |
| Dopamine | Motor control |
| Gamma-aminobutyric acid (GABA) | Motor coordination through inhibition of certain neurons |
| Enkephalins, endorphins | Block transmission and perception of pain |

## Spinal Cord

The **spinal cord** extends through the vertebral canal of the vertebral column to the level of the first lumbar vertebra (L1). It is continuous with the brain through the foramen magnum of the skull. The spinal cord consists of centrally located gray matter involved in reflexes, and peripheral ascending and descending tracts of white matter that conduct nerve impulses to and from the brain. Thirty-one pairs of spinal nerves arise from the spinal cord. The *gray matter* in cross section has a four-horned appearance (Figure 10-1). The **posterior** (*dorsal*) **horns** receive the axons of sensory fibers that enter the spinal cord; the **anterior** (*ventral*) **horns** contain the dendrites and cell bodies of motor neurons that leave the spinal cord. At the thoracic and lumbar level there are **lateral horns** that contain the preganglionic sympathetic neurons that leave via the anterior root. The *white matter* is composed primarily of myelinated fibers that form spinal tracts. The tracts are separated by the horns of gray matter into three regions: **posterior, lateral,** and **anterior funiculi.**

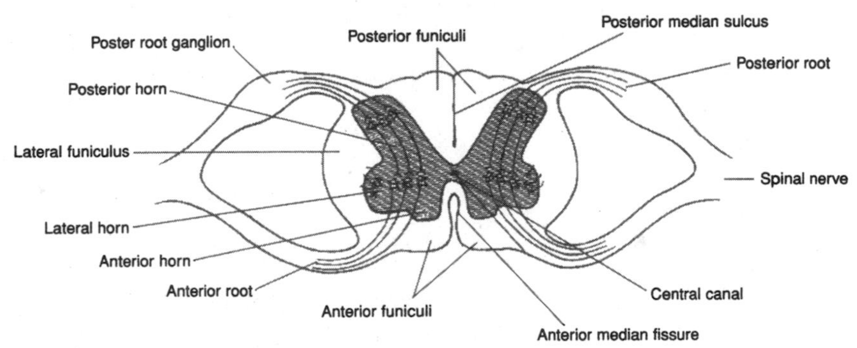


**Figure 10-1.** A cross section of the spinal cord.

## Solved Problems

### True or False

____ 1. The pineal gland, the hypothalmus , and the pituitary gland all have neuroendocrine functions. (**True**)

____ 2. The thalamus is an important relay center in that all sensory impulses (except olfaction) going to the cerebrum synapse there. (**True**)

____ 3. All ventricles of the brain are paired, except for the fourth. (**False**)

____ 4. The posterior horns of the spinal cord contain motor neurons only. (**False**)

**Completion**

1. The ____________ ____________ is the meninx closest to the brain. (**pia mater**)
2. ____________ are neuroglial cells that participate in the blood-brain barrier. (**Astrocytes**)
3. The spinal cord ends at the level of ____________. (**first lumbar vertebra, L1**)

# Chapter 11
# PERIPHERAL AND AUTONOMIC NERVOUS SYSTEMS

IN THIS CHAPTER:

- ✔ *Divisions of the Peripheral Nervous System*
- ✔ *Cranial Nerves*
- ✔ *Spinal Nerves*
- ✔ *Reflex Arc*
- ✔ *Autonomic Nervous System*
- ✔ *Solved Problems*

## Divisions of the Peripheral Nervous System

The **peripheral nervous system** (**PNS**) includes the *cranial nerves*, arising from the inferior aspect of the brain, and the *spinal nerves*, arising from the spinal cord. It is divided into two functional divisions: the **somatic nervous system** and the **autonomic nervous system** (**ANS**).

**Table 11.1** Comparison of Autonomic and Somatic Nervous Systems

| Somatic | Autonomic |
|---|---|
| Conscious or voluntary regulation | Functions without conscious awareness (involuntary) |
| Fibers do not synapse after they leave the CNS (single neuron from CNS to effector organ). | Fibers synapse once at a ganglion after they leave the CNS (two-neuron chain). Motor control. |
| Innervates skeletal muscle fibers, always stimulatory | Innervates smooth muscle, cardiac muscle, and glands; either stimulates or inhibits |

## Cranial Nerves

Cranial nerves innervate structures of the head, neck, and trunk. Most are mixed nerves, some are totally sensory nerves, and others are primarily motor nerves (see Table 11.2). The names of the cranial nerves indicate their primary functions or their general distribution. Cranial nerves are also identified by Roman numerals in order of appearance from front to back.

**Table 11.2** Cranial Nerves, Organized by Function

**Sensory only**

| **Cranial nerve** | **Pathway** | **Function** |
|---|---|---|
| I Olfactory | From olfactory epithelium | Smell |
| II Optic | From retina of eye | Sight |
| VIII Vestibulo-cochlear | From organs of hearing and balance | Hearing; balance and posture |

**Primarily motor** (also proprioceptive function)

| | | |
|---|---|---|
| III Oculomotor | To four eye muscles: *superior*, *inferior*, and *medial recti muscles* and *inferior oblique muscle*; from ciliary body | Movement of eye and eyelid; focusing; change in pupil size; muscle sense |
| IV Trochlear | To (and from) *superior oblique muscle* | Movement of eye; muscle sense |
| VI Abducens | To (and from) *lateral rectus muscle* | Movement of eye; muscle sense |
| XI Accessory | To (and from) neck muscles: *trapezius* and *sternocleidomastoid* | Head and shoulder movement; muscle sense |
| XII Hypoglossal | To (and from) muscles of the tongue | Speech, swallowing; muscle sense |

**Mixed nerves** - carry both sensory and motor fibers

| | | |
|---|---|---|
| V Trigeminal | | |
| $V_1$ Ophthalmic | **Sensory** from cornea, skin of upper 1/3 of face, superior nasal mucosa. | General sensation from skin of face |
| $V_2$ Maxillary | **Sensory** from middle 1/3 of face, teeth and gums of upper jaw, lateral & inferior nasal mucosa. | General sensation from skin of face |
| $V_3$ Mandibular | **Sensory** from lower 1/3 of face, teeth and gums of lower jaw, mucosa of mouth, anterior 2/3 of tongue. | General sensation from skin of face |
| | **Motor** to muscles of mastication. | Chewing of food |

| | | |
|---|---|---|
| VII Facial nerve | **Sensory** from taste buds. **Motor** to facial muscles, lacrimal gland, and salivary glands. | Taste Movement of face; secretion of saliva and tears. |
| IX Glosso-pharyngeal | **Sensory** from pharyngeal muscles and taste buds. **Motor** to pharyngeal muscles and salivary glands | Taste, muscle sense; Swallowing, secretion of saliva |
| X Vagus | **Sensory** from viscera, taste<br><br>**Motor** to viscera | Visceral sensations, taste Visceral muscles and glands |

## Spinal Nerves

There are 31 pairs of spinal nerves: 8 *cervical nerves*, 12 *thoracic nerves*, 5 *lumbar nerves*, 5 *sacral nerves*, and 1 *coccygeal nerve*. The first pair of cervical nerves (C1) emerges between the occipital bone of the skull and the first cervical vertebra (the atlas). The remainder exit the spinal cord and vertebral canal through the *intervertebral foramina*. Each spinal nerve is a *mixed nerve* (carrying both sensory and motor neurons) attached to the spinal cord by a *posterior* (*dorsal*) *root* of sensory fibers and an *anterior* (*ventral*) *root* of motor fibers. A ganglion located on the posterior root, called the *posterior* (*dorsal*) *root ganglion*, contains the cell bodies of sensory neurons. After exiting the vertebral column, the spinal nerve branches into the *anterior* and *posterior rami* (see Figure 11-1).

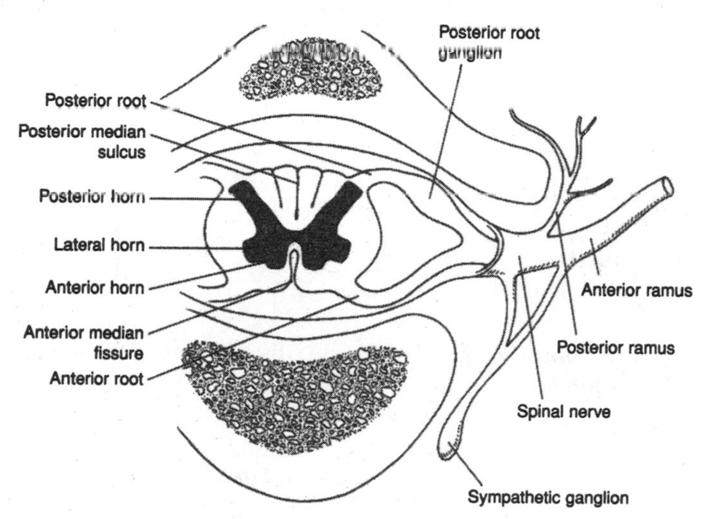


**Figure 11-1.** A cross section of the spinal cord, a spinal nerve, and rami.

The anterior rami of different spinal nerves combine and then split again forming a *plexus*. There are four plexuses of spinal nerves: *the cervical plexus*, *brachial plexus*, *lumbar plexus,* and *sacral plexus* (sometimes referred to together as the *lumbosacral plexus*). Some of the important nerves derived from these plexuses are listed below.

**Table 11.3** Plexus

| Nerve | Plexus | Area of innervation |
|---|---|---|
| **Phrenic** | Cervical | The diaphragm |
| **Musculocutaneous, ulnar, median, axillary, and radial** | Brachial | Muscles of the shoulder and upper limb |
| **Femoral, obturator, saphenous** | Lumbar | Portions of the hip and lower leg |
| **Sciatic** | Sacral | Posterior muscles of the thigh and leg |

## Reflex Arc

There are five components in a typical reflex arc (Figure 11-2):

- **Receptor.** Dendritic endings of a sensory neuron located within the skin, a tendon, joint, or other peripheral organ that responds to specific stimuli.
- **Sensory neuron.** Extending from the receptor through the posterior root, the sensory neuron conveys stimuli to the posterior horn of the spinal cord.
- **Center.** The axon of a sensory neuron synapses with an association neuron within the H-shaped gray matter of the spinal cord.
- **Motor neuron.** Beginning at the synapse with the association neuron, the motor neuron conveys impulses from the anterior horn of the spinal cord, through the anterior root, to the effector organ.
- **Effector.** The muscle or gland that responds to the motor impulse by contracting or secreting, respectively.

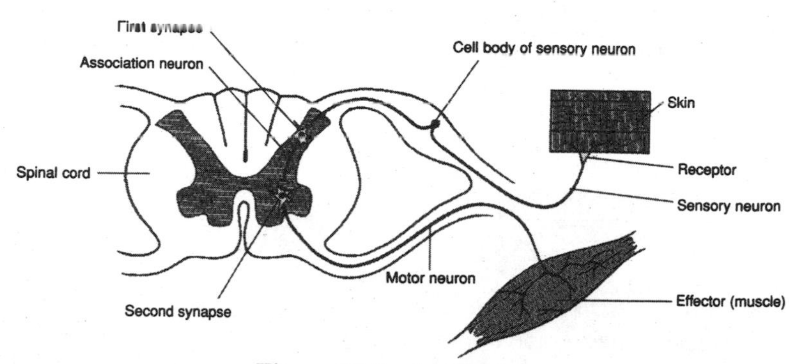


**Figure 11-2.** A reflex arc.

# Autonomic Nervous System

The autonomic nervous system (ANS) functions in maintaining homeostasis. It is divided into two divisions, the *sympathetic* and *parasympathetic divisions*. These are compared below.

| Feature | Sympathetic | Parasympathetic |
|---|---|---|
| Origin of preganglionic fibers | Thoracolumbar nerves | Craniosacral nerves |
| Location of ganglia | Far from visceral effector organs; in sympathetic chain or collateral ganglia | Near or within viscera effector organs |
| Neurotransmitter | In ganglia, acetylcholine; in effector organs, norepinephrine | In ganglia, acetylcholine; in effector organs, acetylcholine |

Sympathetic and parasympathetic fibers are referred to as *adrenergic* and *cholinergic,* respectively, because of the neurotransmitter released at the effector organ. Some sympathetic fibers are cholinergic, those of sweat glands, some vessels in skeletal muscle, the external genitalia, and the adrenal medulla.

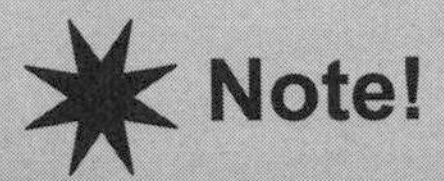


**Receptors for acetylcholine (cholinergic):**

**Muscarinic:** located on effector cells innervated by cholinergic neurons

**Nicotinic:** located at the ganglia of in both divisions of ANS

**Receptors for norepinephrine (adrenergic):**

**α1:** in smooth muscle; effects vasoconstriction, muscular contraction

**α2:** at axon terminal of postganglionic adrenergic neurons; negative feedback: norephinephrin inhibits further its own release

**β1:** in the heart; changes rate and force of contraction.

**β2:** in smooth muscle; effects vasodilation, muscle relaxation

Most of the visceral organs are innervated by both sympathetic and parasympathetic fibers. One division stimulates, while the other inhibits. In general, the effect of sympathetic stimulation is the "Fight or flight" response; the effect of parasympathetic stimulation is the "rest and digest" response. **Sympathetic innervation** increases heart rate and strength of contraction, increases blood pressure, dilates the bronchioles, stimulates sweat glands, increases blood glucose level, decreases digestive activity. **Parasympathetic innervation** decreases heart rate, constricts the bronchioles, increases digestive activity, decreases blood glucose level, stimulates contraction of urinary bladder, dilates the penis.

## Solved Problems

### Completion

1. The ____________ cranial nerve innervates the lateral rectus muscle. (**VI abducens**)
2. The ____________ nerve is a branch of the trigeminal nerve that innervates the lower jaw and teeth, skin of the lower 1/3 of the face, and the tongue. (**mandibular**)
3. There are ____________ cervical nerves, ____________ thoracic nerves, ____________ lumbar nerves, ____________ sacral nerves, and ____________ coccygeal nerve. (**8, 12, 5, 5, 1**)
4. ____________ receptors are located at the ganglia in both sympathetic and parasympathetic divisions of the ANS. (**Nicotinic**)

# Chapter 12
# SENSORY ORGANS

IN THIS CHAPTER:

- ✔ *Taste*
- ✔ *Smell*
- ✔ *Structure and Function of the Eye*
- ✔ *Structure and Function of the Ear*
- ✔ *Solved Problems*

Sensory organs are specialized extensions of the nervous system that contain sensory (afferent) neurons adapted to respond to specific stimuli and conduct nerve impulses to the brain. Sensory organs are very specific as to the stimuli to which they respond.

The senses of the body are classified as **general senses** or **special senses.** General senses include the cutaneous receptors (touch, pressure, heat, cold, and pain) within the skin that provide the *sense of touch.* Special senses are localized in complex receptor organs and have extensive neural pathways. The special senses are the *senses of taste*, *smell*, *sight*, *hearing*, and *balance*.

## Taste

Receptors for the **sense of taste** (*gustation*) are located in taste buds on the surface of the tongue. The taste buds are associated with peglike projections of the tongue called **lingual papillae** (Figure 12-1). A few taste buds are also located in the mucous membranes of the palate and pharynx. A taste bud contains a cluster of 40 to 60 **gustatory cells,** each innervated by a sensory neuron, as well as many more **supporting cells.** The four primary taste sensation are sweet (evoked by sugars, glycols, and aldehydes); sour (evoked by H+, which is why all acids taste sour); bitter (evoked by alkaloids); and salty (evoked by anions of ionizable salts). Sensory innervation of the tongue and pharynx is by a branch of the *facial nerve, CN VI,* from the anterior 2/3 of the tongue, the *glossopharyngeal nerve, CN IX,* from the posterior 1/3 of the tongue, and the *vagus nerve, CN X,* from the pharyngeal region. Taste sensations are transmitted to the *brain stem*, then to the *thalamus*, and finally to the *cerebral cortex*, where taste perception occurs.

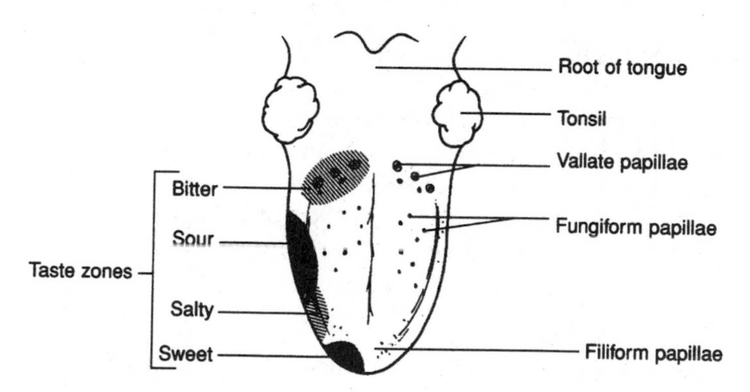


**Figure 12-1.** The surface of the tongue.

## Smell

Receptors for the **sense of smell** (*olfaction*) are located in the nasal mucosa of the superior nasal concha. Like taste receptors, smell receptors are chemoreceptors, specialized neurons that respond to chemical stimuli and require a moist environment to function. The airborne chemicals become dissolved in the mucous layer lining the superolateral part of the nasal cavity. The *olfactory nerve, CN I,* transmits most impulses related to smell. Olfactory sensations are conveyed along each *olfactory tract* to the olfactory portions of the *cerebral cortex* where olfactory perception occurs.

## Structure and Function of the Eye

Accessory structures of the eye either protect the eye or enable eye movement. These structures include the **bony orbit,** the **eyebrow**, the **eyelids,** the **lacrimal apparatus** (**lacrimal glands** that produce lacrimal fluid or tears, and the **lacrimal canals** and **lacrimal sac,** which drain the fluid into the nasal cavity), and the **eye muscles** (responsible for eye movements).

### Muscles Involved in Eye Movements

|  |  |
|---|---|
| **Superior rectus** | rotates the eye superiorly |
| **Medial rectus** | rotates the eye medially |
| **Lateral rectus** | rotates the eye laterally |
| **Inferior rectus** | rotates the eye inferiorly |
| **Superior oblique** | rotates the eye inferolaterally |
| **Inferior oblique** | rotates the eye superolaterally |

**Structure of the Eye**

The spherical eye is approximately 25 mm (1in.) in diameter. It consists of three tunics (layers), a lens, and two principal cavities (Figure 12-2).

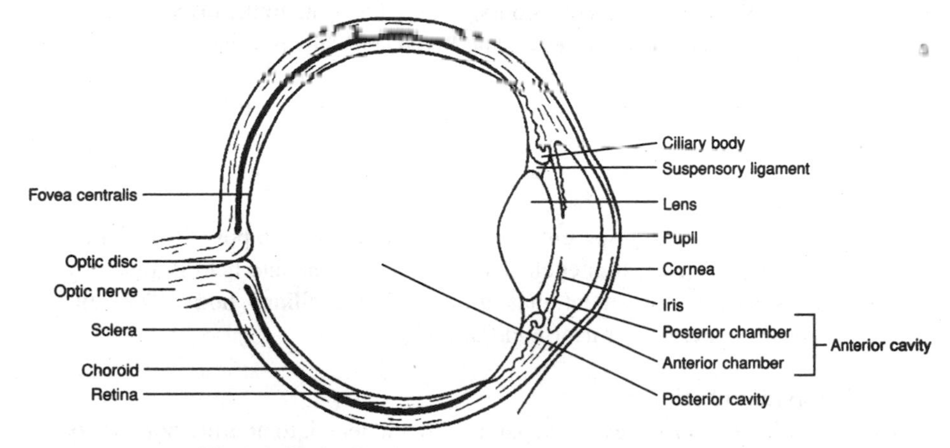


**Figure 12-2.** The internal anatomy of the eye.

**Fibrous Tunic (outer layer)**
The fibrous tunic has two parts. The **sclera** is composed of dense regular connective tissue that supports and protects the eye and is the attachment site for the extrinsic eye muscles. The transparent **cornea** forms the anterior surface of the eye. Its convex shape refracts incoming light rays.

**Vascular Tunic (middle layer)**
The vascular tunic has three parts. The **choroid** is a thin, highly vascular layer that supplies nutrients and oxygen to the eye and absorbs light, preventing it from being reflected. The **ciliary body** is the thickened anterior portion of the vascular tunic. It contains smooth muscle fibers that regulate the shape of the lens. The **iris** forms the most anterior portion of the vascular tunic and consists of pigment (that gives the eye color) and smooth muscle fibers arranged in a circular and radial pattern that regulate the diameter of the **pupil,** which is the opening in the center of the iris.

**Internal Tunic (inner layer,** or **retina)**
The receptor component of the eye contains two types of photoreceptors. **Cones** (approximately 7 million cones per eye) function at high light intensities and are responsible for daytime color vision and acuity (sharpness). **Rods** (approximately 100 million per eye) function at low light intensities and are responsible for night (black-and-white) vision. The retina also contains *bipolar cells*, which synapse with the rods and cones, and *ganglion cells*, which synapse with the bipolar cells. The

axons of the ganglion cells course along the retina to the **optic disc** and form the **optic nerve (CN II)**. The **fovea centralis** is a shallow pit at the back of the retina that contains only cones. It is the area of keenest vision. Surrounding the fovea centralis is the **macula lutea,** which also has an abundance of cones.

### Lens

The lens is a transparent, biconvex structure composed of tightly arranged proteins. It is enclosed in a lens capsule and held in place by the suspensory ligament that attaches to the **ciliary body.** The lens focuses light rays for near and far vision.

### Cavities of the Eye

The interior of the eye is separated by the lens into an **anterior cavity** and a **posterior cavity** (*vitreous chamber*). The anterior cavity is partially subdivided by the iris into an *anterior* and a *posterior chamber*. The anterior cavity contains a watery fluid called *aqueous humor*. The posterior cavity contains a transparent jellylike substance called *vitreous humor*.

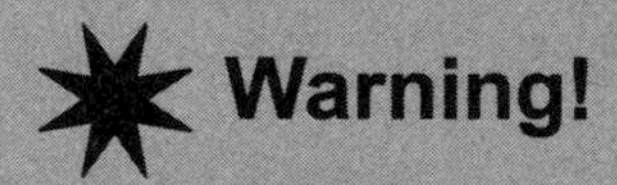


Do **NOT** confuse the **anterior** and **posterior chambers** with the **anterior** and **posterior cavity!** The chambers are subdivisions of the anterior cavity.

### Vision

The *field of vision* is what a person visually perceives. There are three visual fields, the **macular field,** the area of keenest vision, the **binocular field,** the portion viewed by both eyes, but not keenly focused on, and **the monocular field,** that area viewed by one eye and not shared by the other.

The neural pathway for vision consists of the light rays striking the photoreceptors in the retina, which causes the transmission of nerve impulses along the optic nerve to the **optic chiasma.** The optic tract, a continuation of optic nerve fibers from the optic chiasma, carries the impulses to the occipital cerebral lobes where vision occurs.

For an image to be focused on the retina, the more distant the object, the flatter must be the lens. Adjustments in lens shape, accomplished by the ciliary muscles in the ciliary body, are called *accommodation.* When these smooth muscles contract, the fibers within the suspensory ligament slacken, causing the lens to thicken and become more convex.

## Structure and Function of the Ear

The ear is the organ of hearing and equilibrium. It consists of three principal regions: the outer ear, the middle ear, and the inner ear (Figure 12-3).

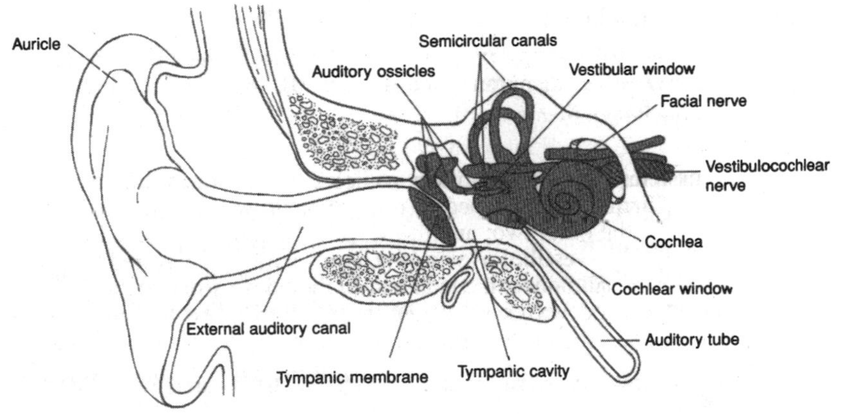


**Figure 12-3.** The ear.

The **outer ear** is open to the external environment and directs sound waves to the middle ear. Structures of the outer ear include the **auricle** (pinna), **the external auditory canal,** and the **tympanic membrane** ("eardrum"). The auricle directs the sound waves to the external auditory canal, a 2.5 cm (1 in.) fleshy tube that fits into the bony external acoustic meatus**.** The thin **tympanic membrane** conducts sound waves to the middle ear

The **middle-ear cavity** or *tympanic cavity* is the air-filled space medial to the tympanic membrane. The structures of the middle ear are the **auditory ossicles,** the **auditory muscles,** and the **auditory (eustachian) tube.** The auditory ossicles are three small bones that extend from the tympanic membrane to the **vestibular (oval) window:** the **malleus** ("hammer"); the **incus** ("anvil"); and the **stapes** ("stirrup"). These small bones amplify the sound waves. The **auditory muscles** are two tiny skeletal muscles that function reflexively to reduce the pressure of loud sounds before it can injure the inner ear. The **auditory (eustachian) tube** connects the middle-ear cavity to the pharynx. It functions to drain moisture from the middle ear cavity and to equalize air pressure on both sides of the tympanic membrane.

The **inner ear** contains the organs of hearing (cochlea) and equilibrium and balance (vestibular apparatus). The structures of the inner ear are described below. **The bony labyrinth** is a network of cavities that consist of three bony **semicircular canals,** the **ampulla** at the base of each semicircular canal, a central **vestibule,** and the **cochlea.** The **membranous labyrinth** is an intercommunicating system of membranous ducts seated in the bony labyrinth. Its parts are co-named with those of the bony labyrinth. The membranous **semicircular canals** and their **ampullae** posses receptors sensitive to rotary motions of the head. The **vestibule** consists of a connecting **utricle** and **saccule,** which possess receptors sensitive to gravity and linear motions of the head. These structures make up the vestibular apparatus. The membranous labyrinth is filled with a fluid, *endolymph*, and to the outside of the membranous labyrinth is a fluid called *perilymph*. The **vestibular (*oval*) window,** a membrane-covered opening from the middle ear into the inner ear, is located at the footplate of the stapes where it transfers sound waves from the solid medium of the auditory ossicles to the fluid medium of the cochlea. Within the cochlea is the membranous **cochlear duct** and the **spiral organ** (*organ of Corti*), the organs of hearing. The **cochlear window** (*round window*) is directly below the vestibular window that reverberates in response to loud sounds.

### Hearing

1. Sound waves are funneled by the auricle into the external auditory meatus.
2. The sound waves strike the tympanic membrane, causing it to vibrate.
3. Vibrations of the tympanic membrane are amplified as they pass through the malleus, incus, and stapes.
4. The vestibular window is pushed back and forth by the stapes setting up pressure waves in the perilymph of the cochlea.
5. The pressure waves are propagated to the endolymph contained within the cochlear duct.
6. Stimulation of the hair cells within the spiral organ of the cochlea causes the generation of nerve impulses in the cochlear nerve (CN VIII), which pass to the pons of the brain.

## Solved Problems

### True or False

____ 1. Taste buds occur on the surface of the tongue, but are also found in smaller number in the mucosa of the palate and pharynx. (**True**)

____ 2. Contraction of the lateral rectus muscle rotates the eye laterally, away from the midline. (**True**)

____ 3. The anterior chamber is located between the cornea and the iris and is filled with vitreous humor. (**True**)

____ 4. The malleus is the bone in the middle ear that is attached to the vestibular window. (**False**)

____ 5. Foramina within the cribriform plate are associated with olfaction. (**True**)

### Matching

Match the structure to its function

| | |
|---|---|
| ____ 1. Cornea (**e**) | (a) provides a sharp visual image |
| ____ 2. Tarsal gland (**h**) | (b) secretes lacrimal fluid (tears) |
| ____ 3. Fovea centralis (**a**) | (c) vibrates in response to sound waves |

| | | |
|---|---|---|
| ____ | 4. Optic radiation (**i**) | (d) attaches to lens capsule |
| ____ | 5. Auditory tube (**g**) | (e) refracts light rays |
| ____ | 6. Lacrimal gland (**b**) | (f) secretes ear wax |
| ____ | 7. Suspensory ligament (**d**) | (g) equalizes air pressure |
| ____ | 8. Ceruminous gland (**f**) | (h) secretes an oily substance |
| ____ | 9. Ciliary body (**j**) | (i) transmits sensory impulses |
| ____ | 10. Basilar membrane (**c**) | (j) secretes aqueous humor |

# Chapter 13
# ENDOCRINE SYSTEM

IN THIS CHAPTER:

- ✔ *Hormones*
- ✔ *Negative and Positive Feedback*
- ✔ *Endocrine Glands and Their Secretions*
- ✔ *Solved Problems*

The **endocrine system** consists of *endocrine glands* that secrete specific chemicals called *hormones* into the blood or surrounding interstitial fluid. The endocrine system functions closely with the nervous system in regulating and integrating body processes. More specifically, hormones cause changes in the metabolic activities in specific cells, and nerve impulses cause muscles to contract or glands to secrete. In general, the action of hormones is relatively slow and the effects are prolonged, whereas the action of nerve impulses is fast and the effects are of short duration.

## Hormones

A **hormone** is a chemical messenger secreted by an endocrine gland. Its chemical composition is such that it has its effect on specific receptor sites on target cells causing a series of biochemical events that leads to a specific response. Hormones are classified according to chemical structure and the location of the cell membrane receptors on their target cells.

**Table 13.1** Classes and Composition of Hormones

| Classes of hormones | Composition |
|---|---|
| **Amino acids derivative (catecholamines)** | C, H, and N, amine ($NH_2$) group |
| **Polypeptides** | Long chains of amino acids |
| **Glycoproteins** | Large proteins combined with carbohydrates |
| **Steroids** | Lipids |
| **Fatty acid derivatives** | Long hydrocarbon acid chains. |

Hormones are divided into two groups, based on the location of the receptors on the target cells. **Group I hormones** bind to intracellular receptors and are lipophilic. These include the steroid hormones. **Group II hormones** bind to cell surface receptors and are hydrophilic. These hormones include polypeptide, glycoprotein, and catecholamine hormones.

## Negative and Positive Feedback

**Negative feedback** involves a chain of biochemical or physiological events. Generally, an increased amount of end product inhibits the production, mechanism, or action of a starting substance to prohibit further synthesis of the end product. For example: $A > B > C > D$. As $A$ progresses through $B$ and $C$ to $D$ the amount of $D$ increases. Substance $D$, however, is an inhibitor of substance $A$. As levels of substance $D$

increase, substance *A* receives "negative feedback" to prevent the process from continuing to produce more *D*. An example of this mechanism is the production of cortisol and its regulation along the hypothalamic-pituitary-adrenal axis.

In the case of **positive feedback,** *D*, in the example above, would stimulate A to further produce increased amounts of *B*, and so on to *D*. This is less common in the body. One example is the secretion of oxytocin stimulating contraction of uterine muscles during labor. As labor progresses pressure on the cervix continues to stimulate the release of oxytocin, which continues to stimulate uterine muscles to contract.

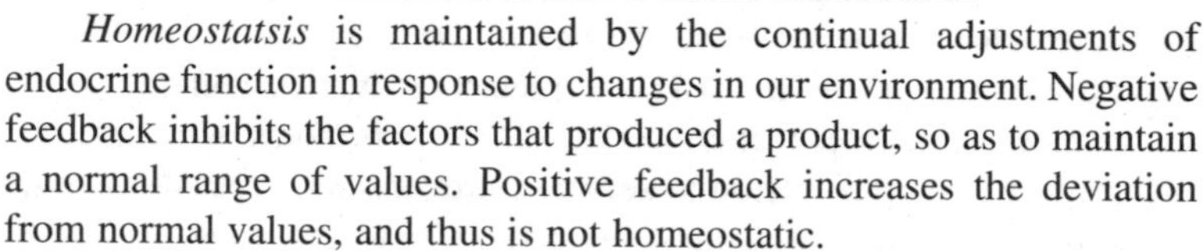

*Homeostatsis* is maintained by the continual adjustments of endocrine function in response to changes in our environment. Negative feedback inhibits the factors that produced a product, so as to maintain a normal range of values. Positive feedback increases the deviation from normal values, and thus is not homeostatic.

## Endocrine Glands and Their Secretions

Endocrine organs are widely scattered throughout the body with no anatomical continuity. In addition to the discrete endocrine organs, several other organs, referred to as *mixed organs*, have an endocrine function. These include the thymus, stomach, duodenum, placenta and the heart.

### Pituitary gland

Location and structure: Located on the inferior side of the brain in the sella turcica of the sphenoid bone. The *pituitary stalk* attaches the gland to the hypothalamus portion of the brain. The pituitary gland is divided into an *anterior lobe*, the **adenohypophysis,** and a *posterior lobe*, the **neurohypophysis.**

Secretions and effects:
Secreted by the adenohypophysis:

- **Human growth hormone (HGH).** Targets the bones and soft tissues. Accelerates rate of body growth by 1) stimulating amino acid uptake by cells; 2) increasing synthesis of tRNA; and 3) increasing the number and aggregation of ribosomes, thus promoting protein synthesis.

- **Thyroid-stimulating hormone (TSH).** Targets the thyroid gland. Stimulates synthesis and release of thyroid hormones.
- **Adrenocorticotropic hormone (ACTH).** Targets the adrenal cortex. Stimulates secretion of glucocorticoids.
- **Prolactin (PRL).** Targets the mammary glands. Promotes development of mammary glands and stimulates milk production. Regulated by production of placental hormones during pregnancy and stimulation of the nipple during lactation.
- **Follicle stimulating hormone (FSH).** Targets the ovaries and testes. Stimulates growth of ovarian follicles and spermatogenesis in females and males respectively.
- **Luteinizing hormone (LH).** Targets the ovaries and testes. In females, stimulates maturation of follicles, promotes ovulation and stimulates corpus luteum to secrete estrogens and progesterone. In males, stimulates interstitial cells to secrete testosterone.

Secreted by the neurohypophysis:

- **Antidiuretic hormone (ADH).** Targets the kidney tubules. Facilitates water reabsorption in the distal convoluted tubules and collecting ducts. Release stimulated by dehydration and increased plasma osmolarity. Controlled by negative feedback.
- **Oxytocin.** Targets the uterus and mammary glands. Stimulates contraction of uterine muscles and secretion of milk. Release stimulated by stretching of uterus late in pregnancy and by mechanical stimulation of the nipple during nursing. Controlled by positive feedback.

**Thyroid gland**

Location and structure: Located in the neck on either side of the thyroid cartilage at the top of the trachea.

Secretions and effects: Thyroid hormones **triodothyronine** (**T3**) and **tetraiodotyrosine** (**T4** or **thyroxine**) are secreted under the stimulation of TSH from the pituitary. These hormones accelerate metabolic rate, oxygen consumption, and glucose obsorption; increase body temperature; affect growth and development; and enhance the effects of the sympathetic nervous system.

**Parathyroid glands**

Location and structure: Small glands embedded in the posterior surface of the thyroid gland.

Secretion and effects: **Parathyroid hormone (PTH)** increases plasma calcium levels by 1) stimulating the formation and activity of osteoclasts, which break down bone tissue and release calcium from the bones into the blood; 2) acting on kidney tubules to increase calcium reabsorption; and 3) increasing synthesis of 1/25-dihydroxycholecalciferol, which increases calcium absorption from the GI tract. Release stimulated by decreased plasma calcium concentration.

**Adrenal glands**

Location and structure: Triangular shaped gland embedded in adipose tissue at the superior borders of the kidneys. Consists of an outer **adrenal cortex** and an inner **adrenal medulla.**

Secretions and effects:

Secreted by the adrenal cortex:

- **Glucocorticoids (corticosterone, cortisol).** 1) Regulate carbohydrate and lipid metabolism, accelerate breakdown of proteins; 2) in large doses, inhibit inflammatory responses; 3) promote vasoconstriction; and 4) help the body resist stress. Secretion controlled by ACTH from the anterior pituitary and negative feedback mechanisms.
- **Mineralocorticoids (deoxycorticosterone, aldosterone).** Regulate concentration of extracellular electrolytes, especially sodium and potassium. Aldosterone secretion controlled by the reninangiotensin system (Figure 13-1), plasma K+ concentration, and ACTH.

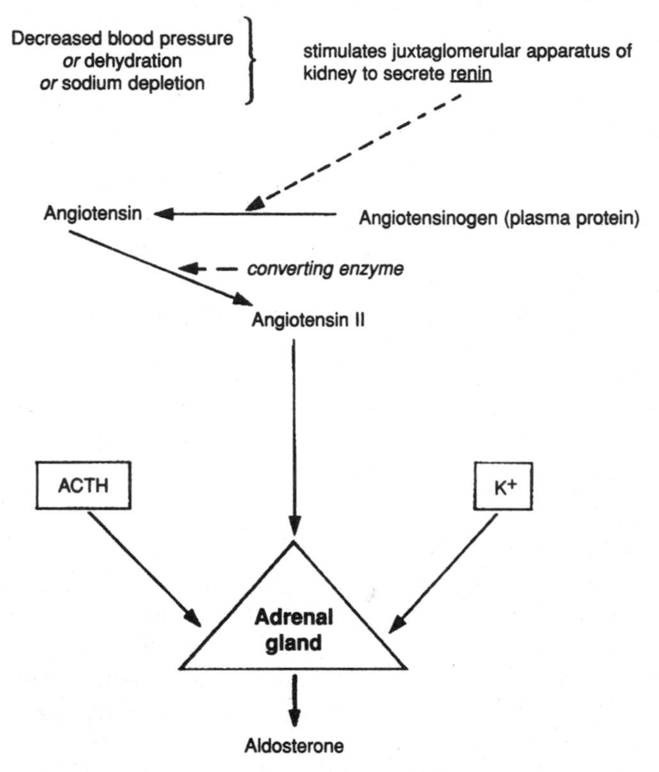


**Figure 13-1.** The sequence of events in aldosterone production.

Secreted by the adrenal medulla:

- **Amine hormones** (**epinephrine** and **norepinephrine**). Results in sympathetic response (see Chap. 11). Increases release of ACTH and TSH from the adenohypophysis. Release stimulated by sympathetic stimulation.

**Pancreas**

Location and structure: In the abdomen, inferior to the stomach. Endocrine portion consists of scattered clusters of cells called **pancreatic islets** (*islets of Langerhans*).

<u>Secretions and effects</u>:

- **Glugagon** is secreted by alpha cells innervated by cholinergic fibers. It stimulates glycogenolysis and maintains blood glucose levels during fasting or starvation.
- **Insulin** is secreted by beta cells innervated by adrenergic fibers. It stimulates movement of blood glucose across cells, stimulates glycolysis and lowers blood glucose levels.
- **Somatostatin,** secreted by delta cells, stimulates incorporation of sulfur into cartilage and stimulates collagen formation.

**Endocrine secretions from mixed glands**

- **Thymus**. Secretes thymosin; stimulates T-lymphocyte activity.
- **Pineal gland.** Secretes melatonin; affects secretion of gonadotropins and ACTH from the anterior pituitary.
- **Gastric mucosa.** G cells secrete gastrin; stimulates gastric juice secretion and gastric motility.
- **Duodenal mucosa.** Secretes secretin; stimulates secretion of pancreatic juice.
- **Placenta.** Secretes human chorionic gonadotropin (hCG), human somatomammatropin (hCS), estrogens, and progesterone.

**You Need to Know** 

- **Name, location,** and **secretion** of each endocrine organ.
- **Target tissue** and **effect** for each endocrine secretion.

## Solved Problems

**True or False**

____ 1. Two hormones are never present in the blood at the same time. (**False**)

____ 2. The cells of the parathyroid gland respond directly to the glucose concentration in the blood. (**False**)

____ 3. The posterior pituitary is not composed of true glandular tissue. (**False**)

# Chapter 14
# CARDIOVASCULAR SYSTEM: BLOOD

IN THIS CHAPTER:



## Functions of Blood

Blood is a fluid connective tissue that is pumped by the heart through the vessels (arteries, arterioles, capillaries, venules, and veins) of the cardiovascular system.

- **Transports** oxygen, nutrients, and hormones to body tissues, and carbon dioxide and waste materials from tissues to be excreted.
- **Acid-base regulation.** Controls respiratory acidosis (low pH) or alkalosis (high pH) through the bicarbonate buffer system. H+

combines with bicarbonate to form carbonic acid, which dissociates to form $CO_2$ and $H_2O$. $CO_2$ is exhaled, and blood becomes less acidic.

- **Thermoregulation.** During hyperthermia, carries excess heat to the body surface.
- **Immunity.** Leukocytes (white blood cells) are transported to sites of injury or invasion by disease-causing agents.
- **Hemostasis.** Thrombocytes (platelets) and clotting proteins minimize blood loss when a blood vessel is damaged.

## Composition of Blood

Blood is composed of a liquid matrix (**blood plasma**) and several types of formed elements (**red blood cells, white blood cells,** and **platelets**) (see Figure 14-1). The plasma contains a variety of proteins and many other small molecules and ions. Blood minus the formed elements and the clotting proteins is called **serum.**

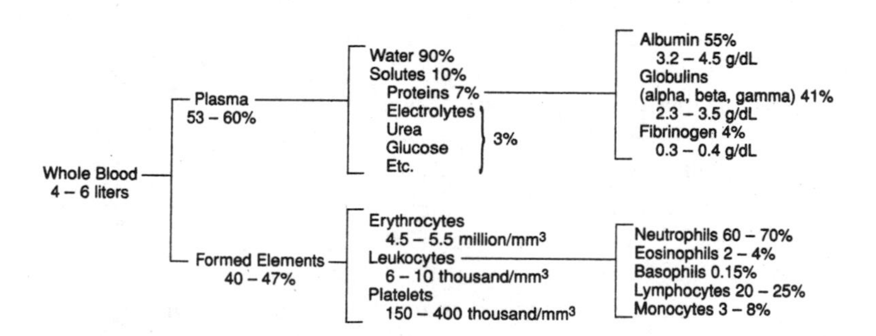


**Figure 14-1.** The composition of blood.

## Erythrocytes

An **erythrocyte,** or red blood cell (RBC), is a flexible, biconcave, anucleated cell. During embryonic development, *erthropoiesis* (the manufacture of red blood cells) occurs first in the yolk sac. Production then moves to the liver, spleen, and red bone marrow. The main constituent of RBCs is **hemoglobin** and the essential function is to bring oxygen, carried by hemoglobin, to all parts of the body. The **hematocrit** is the percentage of total blood volume occupied my erythrocytes.

**Erythropoiesis** is the manufacture of RBC. The sequence of cellular differentiation is as follows:

hemocytoblast → proerythroblast → erythroblast → normoblast → reticulocyte → erythrocyte

The substances required for erythrocyte production include: proteins, lipids, amino acids, iron, Vitamin $B_{12}$, folic acid, copper, and cobalt.

RBCs have an average lifespan of 120 days. New RBCs are formed in the bone marrow and old RBCs are destroyed in the liver and spleen. Any condition that decreases oxygen in the body tissues will, by negative feedback, increase erythropoiesis. In response to low oxygen concentration, the kidneys secrete the hormone *erythropoietin,* which stimulates erythropoiesis in the bone marrow.

The **hemoglobin** (Hb) molecule contains four iron containing molecules (heme) and four polypeptide chains (globin). Each heme portion can bind four molecules of oxygen. Each RBC has approximately 280 million hemoglobin molecules, and can thus carry over a billion molecules of oxygen. Hemoglobin can also bind $CO_2$ and CO (carbon monoxide) molecules. $CO_2$ and $O_2$ have distinct carry sites on the Hb molecule. CO binds to a heme at the same site as $O_2$ and has a greater affinity for the heme, thus excluding $O_2$ binding at that site. This exclusion of $O_2$ makes CO such a dangerous gas. The breakdown of erythrocytes in the liver and spleen results in molecules used to form *bile*, a digestive secretion formed in the liver. These by-products are excreted in feces or urine, giving them their distinctive brown or yellow color.

## Platelets

**Platelets,** or *thrombocytes*, are small cellular fragments that originate in the bone marrow from a giant cell, a **megakaryocyte**. Megakaryocytes form platelets by pinching off bits of cytoplasm and extruding them into the blood. Platelets contain several clotting factors, calcium ions, ADP, serotonin, and various enzymes; they play an important role in **hemostasis** (the arrest of bleeding).

The major events in **hemostasis** are:

1. *constriction of the blood vessels*;
2. *plugging the wound by aggregated platelets*; and
3. *clotting of the blood into a mass of fibrin* that augments the plug in sealing the wound and providing a framework for repair.

In the event of a vessel defect or injury, platelets aggregate to form a plug. Adenosin diphosphate (ADP) and thromboxane A2 released from the platelets further enhance platelet aggregation. The platelet plug aids in reducing blood loss at the site by:

1. physically sealing the vessel defect,
2. releasing chemicals that cause vasoconstriction, and
3. releasing other chemicals that stimulate blood clotting. There are multiple clotting factors produced in the liver involved in clotting.

## Leukocytes

There are five types of leukocytes (white blood cells).

**Table 14.1** Type Description and Function of Leukocytes

| Type | Description | Function |
|---|---|---|
| **Neutrophils** | Lobed nucleus, fine granules | Phagocytosis |
| **Eosinophils** | Lobed nucleus, red or yellow granules | Phagocytize antigen-antibody complexes |
| **Basophils** | Obscure nucleus, large purple granules | Release heparin, histamine, and serotonin |
| **Lymphocytes** (B cells and T cells) | Round nucleus, little cytoplasm | Produce antibodies, destroy specific target cells |
| **Monocytes** | Kidney shaped nucleus | Phagocytosis |

## Blood Plasma

Blood plasma is composed of the following:

- **Water**
- **Proteins** (albumins, globulins, and fibrinogens)
  **Electrolytes** ($Na^+$, $K^+$, $Ca^{2+}$, $Mg^{2+}$, $Cl^-$, $HCO_3^-$, $HPO_4^{2-}$, $SO_4^{2-}$)
- **Nutrients** (glucose, amino acids, lipids, cholesterol, vitamins, trace elements)
- **Hormones**
- **Dissolved gasses** ($CO_2$, $O_2$, N)
- **Waste products** (urea, uric acid, creatinine, bilirubin)

**Albumins,** the smallest and most abundant proteins in the blood, maintain the osmotic pressure of the blood, buffer the blood, and contribute to the viscosity of blood. The **globulin** proteins in the blood function in transport, enzymatic action, clotting, and immunity. The **electrolytes** are necessary for membrane transport, blood osmolarity, and neurological function.

## Solved Problems

### True and False

____ 1. Blood functions in transport, pH balance, thermoregulation, and immunity mechanisms. (**True**)

____ 2. The major mechanisms of homeostasis are plugging, clotting, and constriction. (**True**)

____ 3. Erythrocyte production requires folic acid, copper, protein, polysaccharides, and biliverdin. (**False**)

### Matching

____ 1. Thrombocyte (**e**)

____ 2. Eosinophil (**f**)

____ 3. Neutrophil (**d**)

____ 4. Leukocyte (**g**)

(a) biconcave, anucleated cell

(b) selective defender against invaders

(c) obscure nucleus; stains with large purple granules

(d) lobed nucleus and fine granules; stains with neutral dyes

| | |
|---|---|
| ____ 5. Lymphocyte **(b)** | (e) formation of clots |
| ____ 6. Basophil **(c)** | (f) granules that take up the red dye eosin |
| ____ 7. Erythrocyte **(a)** | (g) white blood cell |

# Chapter 15
# CARDIOVASCULAR SYSTEM: THE HEART

IN THIS CHAPTER:

- ✔ *Structure*
- ✔ *Blood Flow through the Heart*
- ✔ *Fetal Circulation*
- ✔ *Coronary Circulation*
- ✔ *Conduction System and Innervation*
- ✔ *Cardiac Cycle*
- ✔ *Electrocardiogram*
- ✔ *Solved Problems*

## Structure

The **heart** is a hollow, four-chambered muscular organ that is specialized for pumping blood through the vessels of the body. It is located in the *mediastinum* where it is surrounded by a tough fibrous membrane called the **pericardium.** The **parietal pericardium** is a loose sac composed of an outer *fibrous layer* that protects the heart and an inner *serous layer* that secretes **pericardial fluid.** The **visceral pericardium** is a serous membrane that makes up the outer layer of the wall of the heart (the **epicardium**). The space between the parietal pericardium and the visceral pericardium is the **pericardial cavity.** Pericardial fluid is found within this cavity and functions to lubricates the surface of the heart.

The heart is composed of three layers from superficial to deep:

- **Epicardium:** Serous membrane of connective tissues covered with epithelium
  Outer coating that lubricates
- **Myocardium:** Cardiac muscle tissue and connective tissues
  Contractile layer, thickest layer
- **Endocardium:** Epithelial membrane and connective tissues
  Strengthened protective inner lining of heart

**Internal structure**

The heart is a four-chambered double pump (Figure 15-1). It consists of upper right and left **atria** that pulse together, and lower right and left **ventricles** that also contract together. The atria are separated by the thin, muscular **interatrial septum,** while the ventricles are separated by the thick, muscular **interventricular septum. Atrioventricular valves** (AV valves) are located between the upper and lower chambers of the heart, and the **semilunar valves** are at the bases of the two large vessels (the pulmonary trunk and aorta) that leave the heart.

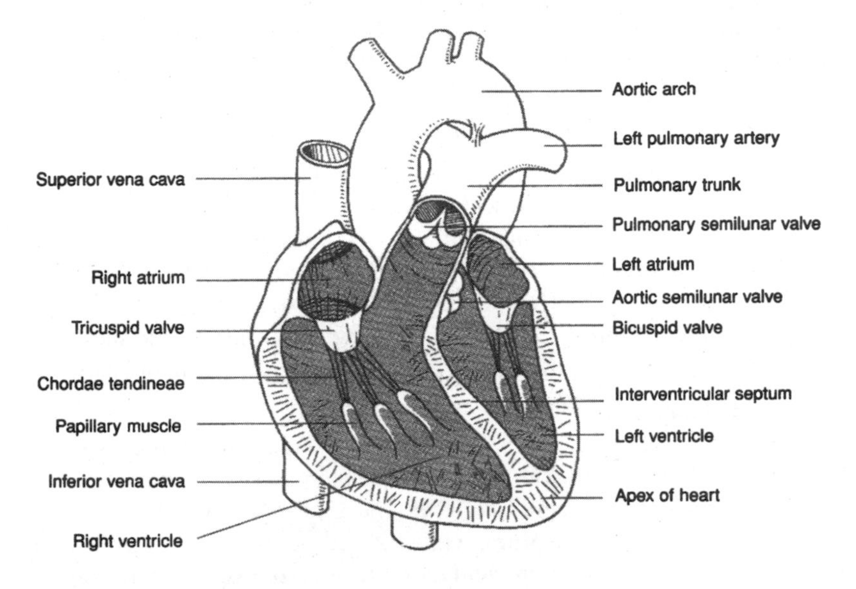


**Figure 15-1.** Internal anatomy of the heart.

## Heart Valves

**Tricuspid valve:** Between the right atrium and the right ventricle

**Bicuspid valve:** Between the left atrium and the left ventricle

**Pulmonary semilunar valve:** Between right ventricle and pulmonary trunk

**Aortic semilunar valve:** Between left ventricle and ascending aorta

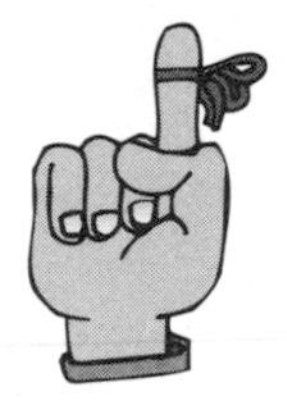

Each cusp of the atrioventricular valves is held in position by strong tendinous cords, the **chordae tendineae,** which are secured to the ventricular wall by cone-shaped **papillary muscles.** All the valves in the heart prevent the backflow of blood to the previous chamber upon contraction of the muscular walls of the heart.

## Blood Flow through the Heart

Deoxygenated blood returning from the body fills the right atrium while oxygenated blood returning from the lungs fills the left atrium. Blood flows from the atria into the ventricles, the atria contract, emptying the remaining blood into the ventricles. The ventricles then contract, forcing blood from the right ventricle into the pulmonary trunk and from the left ventricle into the aorta.

The **pulmonary circuit** refers to the circulation of blood from the heart to the lungs and then back to the heart. The structures that are part of the pulmonary circuit are the right ventricle, pulmonary trunk and pulmonary arteries, the capillary network in the lungs, the pulmonary veins returning blood to the heart, and the left atrium, which receives this oxygenated blood. The **systemic circuit** refers to the circulation of blood to and from all of the other body tissues. The components of this circuit are the left ventricle, the arteries, capillaries, and veins going to all the body tissues, and the right atrium, which receives deoxygenated blood as it returns from all the body tissues.

## Fetal Circulation

In a fetus, the lungs are nonfunctional and oxygen and nutrients are obtained from the placenta (Figure 15-2). An umbilical cord connects the fetus to the placenta. The umbilical cord consists of an **umbilical vein** that transports oxygenated blood toward the heart and two **umbilical arteries** that return deoxygenated blood to the placenta. Three fetal structures reroute blood in the fetus:

| Fetal structure | Function | Adult remnant |
|---|---|---|
| **Ductus venosus** | Allows blood to bypass the liver | Ligamentum venosum |
| **Foramen ovale** | Shunts blood directly from the right atrium to the left atrium | Fossa ovalis |
| **Ductus arteriosus** | Shunts blood from the pulmonary trunk to the aortic arch | Ligamentum arteriosum |

The umbilical vein forms the *round ligament* of the liver and the umbilical arteries become the lateral *umbilical ligaments* in an adult.

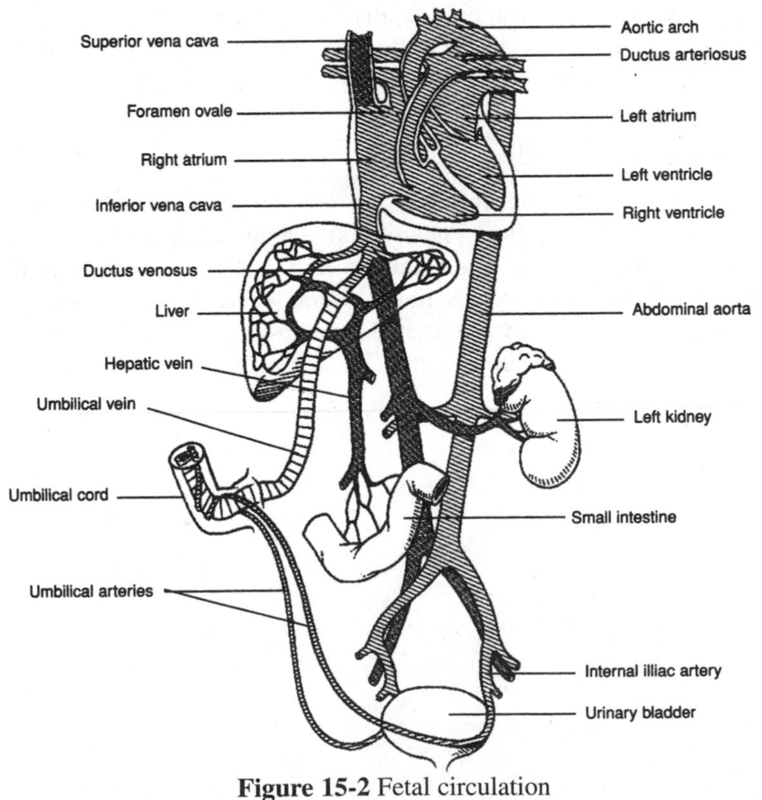


**Figure 15-2** Fetal circulation

## Coronary Circulation

Blood supply to the myocardium is from the **right** and **left coronary arteries,** which exit the ascending aorta just beyond the aortic semilunar valve. The left coronary artery gives rise to the **anterior interventricular** and **circumflex arteries,** and the right coronary artery gives rise to the **posterior interventricular** and **marginal arteries.** The **great cardiac vein** and **middle cardiac vein** return blood to the **coronary sinus,** which empties into the right atrium.

## Conduction System and Innervation

The **conduction system** consists of *nodal tissues* (specialized cardiac muscle fibers) that initiate the conduction of depolarization waves through the myocardium. The pacemaker of the heart is the **sinoatrial node** (SA node) located in the posterior wall of the right atrium. It depolarizes spontaneously at the rate of 70 to 80 times per minute, causing the atria to contract. Impulses from the SA node pass to the **atrioventricular node** (AV node) in the interatrial septum, the **atrioventricular bundle** (AV bundle) in the interventricular septum, and finally to the **conduction myofibers** (Purkinje fibers) within the ventricular walls. Stimulation of the conduction myofibers causes the ventricle to contract simultaneously. The SA and AV nodes are innervated by sympathetic and parasympathetic nerve fibers. Sympathetic impulses accelerate heart action; parasympathetic impulses through the vagus (CN X) decelerate heart action. These impulses are regulated by the cardiac centers in the hypothalamus and medulla oblongata.

## Cardiac Cycle

The cardiac cycle consists of a phase of relaxation, called **diastole,** followed by a phase of contraction, referred to as **systole.** Major events of the cycle, starting in mid-diastole, are as follows:

- **Late diastole.** The atria and ventricles are relaxed, the AV valves are open, and the semilunar valves are closed. Blood flows passively from the atria into the ventricles.
- **Atrial diastole.** The atria contract and pump the additional blood into the ventricles.

- **Ventricular systole.** At the beginning of ventricular contraction, the AV valves close, causing the first heart sound, "lub." When pressure in the right ventricle exceeds the diastolic pressure in the pulmonary artery (10 mmHg) and the left ventricular pressure exceeds diastolic pressure in the aorta (80 mmHg) the semilunar valves open and ventricular ejection begins. Under normal resting conditions, the pressure reaches 25 mmHg on the right side and 120 mmHg on the left side. The stroke volume, volume of blood ejected from either ventricle, is 70 to 90 mL.
- **Early diastole.** As the ventricles begin to relax, the pressure drops rapidly. The semilunar valves close, preventing backflow into the ventricles from the arteries and causing the second heart sound, "dub." The AV valves open and blood begins to flow from the atria into the venticles.

**Cardiac output,** the volume of blood pumped by the left ventricle in 1 minute, may be calculated as:

Cardiac output (C.O.) = stroke volume (S.V.) × heart rate (H.R.)

**C.O. is increased** by:

1. sympathetic stimulation of the heart.
2. increased end-diastolic volume (*Starling's law of the heart*).
3. various forms of anemia that result in decreased total peripheral resistance.

**C.O. is decreased** by decreased venous return.

## Electrocardiogram

Because the body is a good conductor of electricity, potential differences generated by the depolarization and repolarization of the myocardium can be detected on the surface of the body and recorded as an **electrocardiogram** (**ECG** or **EKG**) (Figure 15-3). The *P wave* indicates depolarization of the atria. The *QRS* complex is the record of ventricular depolarization; the *T wave*, of ventricular repolarization. The short flat segment between S and T represents the refractory state of the ventricular myocardium; that between P and Q a nonconductive phase of the AV node, during which atrial systole can be completed.

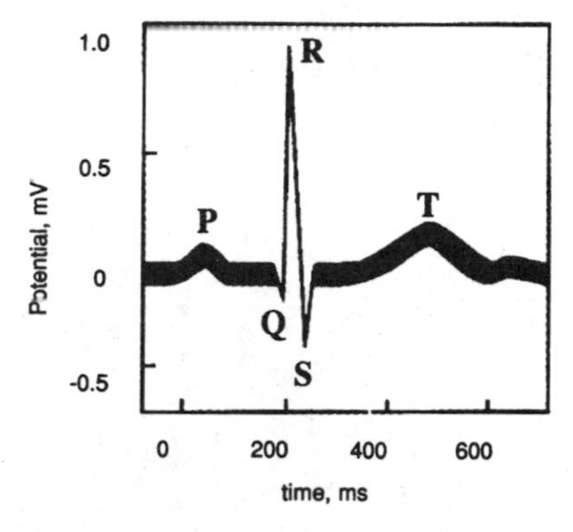


**Figure 15-3.** A normal electrocardiogram (ECG).

## Solved Problems

| | | | |
|---|---|---|---|
| ____ 1. | P wave (**a**) | (a) | atrial depolarization |
| ____ 2. | First heart sound (**e**) | (b) | cardiac output |
| ____ 3. | QRS complex (**c**) | (c) | ventricular depolarization |
| ____ 4. | S.V. × H.R. (**b**) | (d) | ventricular repolarization |
| ____ 5. | T wave (**d**) | (e) | closure of the AV valves at the onset of systole |

# Chapter 16
# Cardiovascular System: Vessels and Blood

In This Chapter:

- ✔ *Vessels—Arteries, Capillaries, and Veins*
- ✔ *Principal Systemic Arteries*
- ✔ *Principal Systemic Veins*
- ✔ *Blood Pressure*
- ✔ *Solved Problems*

The functions of the cardiovascular system are similar to the functions of blood listed in Chapter Fourteen.

## Vessels—Arteries, Capillaries, and Veins

The walls of blood vessels are composed of the following tunics (layers): the **tunica interna,** an inner layer of squamous epithelium, called *endothelium,* resting on a layer of connective tissue; the **tunica media,** a middle layer of smooth muscle fibers mixed with elastic fibers; and the **tunica externa,** an outer layer of connective tissue containing elastic and collagenous fibers. The tunica externa of the larger vessels is infiltrated with a system of tiny blood vessels called the *vasa vasorum* ("vessels of the vessels") that nourish the more external tissues of the blood vessel wall.

**Table 16.1** Structure and Function of Vessels

| **Vessel** | **Structure** | **Function** |
|---|---|---|
| **Artery** - carries blood away from the heart | Strong elastic vessel, contains all three tunics; lumen diameter large relative to wall thickness | Distributive channel to body tissues; blood carried under high pressure |
| **Arteriole** - branches off of small arteries | Thick layer of smooth muscle in tunica media; relatively narrow lumen | Alters diameter to control blood flow, dampen pulsate flow to a steady flow |
| **Capillary** - exchange area of the system | Wall composed of single layer of endothelium (tunica interna); smooth muscle cuff at its origin regulates flow | Allow exchange of fluids, nutrients, and gases between the blood and the interstitial fluids |
| **Venules** (small veins) and **Veins** - carry blood back to the heart | Thin, distensible vessel consisting of three tunics; lumen diameter very large; **valves** present | Serves as fluid reservoir (hold 60% to 75% of blood volume); constricts in response to sympathetic stimulation; valves ensure unidirectional blood flow |

## Principal Systemic Arteries

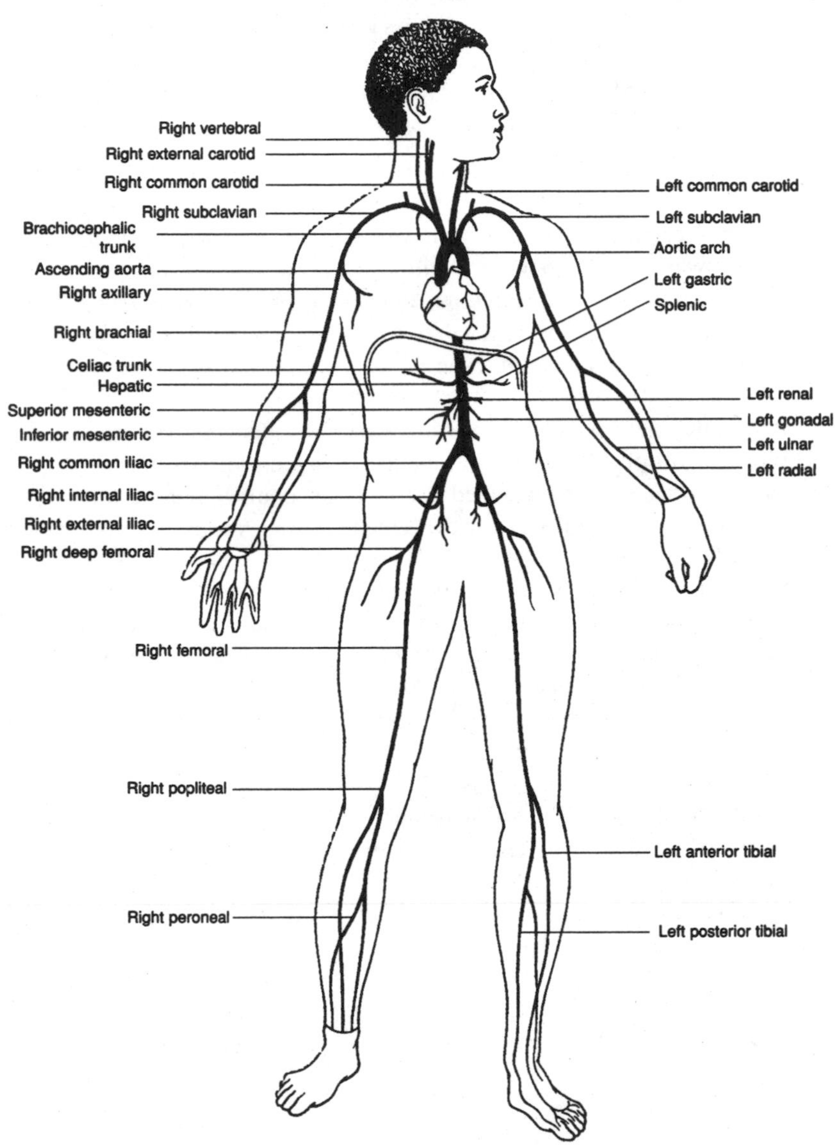


**Figure 16-1.** Principal arteries of the body.

Arteries to the head, neck, and upper limbs:

Three vessels branch off of the **aortic arch,** the **brachiocepahlic trunk,** the **left common carotid a.,** and the **left subclavian a.** The brachiocephalic trunk branches into a **right common carotid a.** and a **right subclavian a.** Branches off these vessels supply the head, neck, and shoulder regions. The right and left subclavian arteries continue into the upper limb as the **axillary a.,** which then becomes the **brachial a.** Branches off these vessels supply the upper limb.

Paired arteries off the thoracic aorta:

| Artery | Region or organ served |
|---|---|
| Pericardial arteries | Pericardium |
| Intercostal arteries | Thoracic wall |
| Bronchial arteries | Right and left bronchus |
| Esophageal arteries | Esophagus |
| Superior phrenic arteries | Diaphragm |

Arteries from the abdominal aorta:

| | |
|---|---|
| Inferior phrenic arteries | Diaphragm |
| Celiac trunk | |
| Hepatic artery | Liver, upper pancreas, duodenum |
| Splenic artery | Spleen, pancreas, stomach |
| Left gastric artery | Stomach, esophagus |
| Superior mesenteric artery | Small intestine, pancreas, cecum, appendix, ascending colon, transverse colon |
| Suprarenal arteries | Adrenal (suprarenal) glands |
| Renal arteries | Kidneys |
| Gonadal arteries | Gonads (testes; ovaries) |
| Inferior mesenteric artery | Transverse colon, descending colon, sigmoid colon, rectum |
| Common iliac arteries | |
| External iliac arteries | Lower extremities |
| Internal iliac arteries | Reproductive organs, gluteal muscles |

## Principal Systemic Veins

The major veins that return blood to the heart are the **superior vena cava**, returning blood from the head, neck and upper extremity, and **inferior vena cava**, carrying blood from the abdomen and lower extremity. The major veins are diagramed in Figure 16-2.

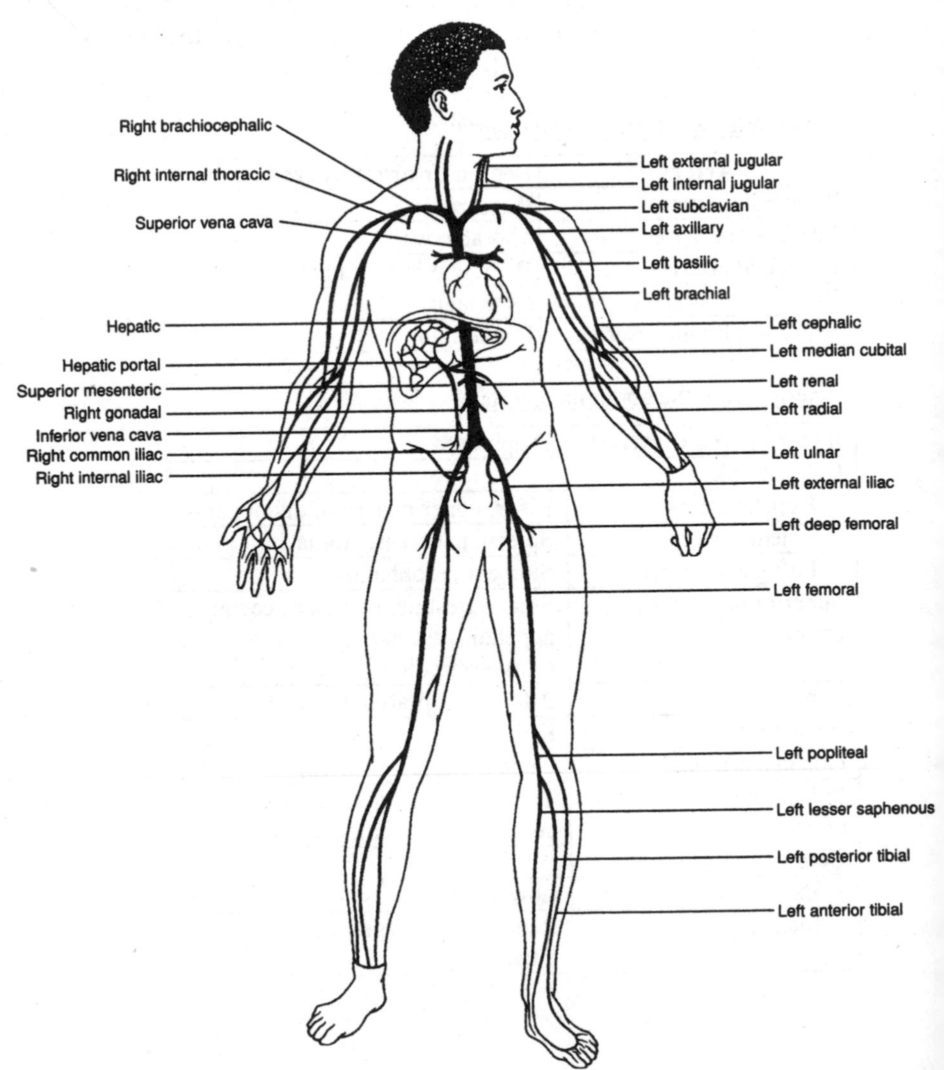


**Figure 16-2** Principal veins of the body.

## Blood Pressure

Blood pressure is the force per unit area exerted by the blood against the inner walls of blood vessels; it is due primarily to the action of the heart.

Factors that affect blood pressure:

- **Heart rate:** increased rate increases pressure
- **Blood volume:** increased volume increases pressure
- **Peripheral resistance:** decreased vessel diameter, increases resistance and thus pressure.

Normal blood pressure is about 120/80.

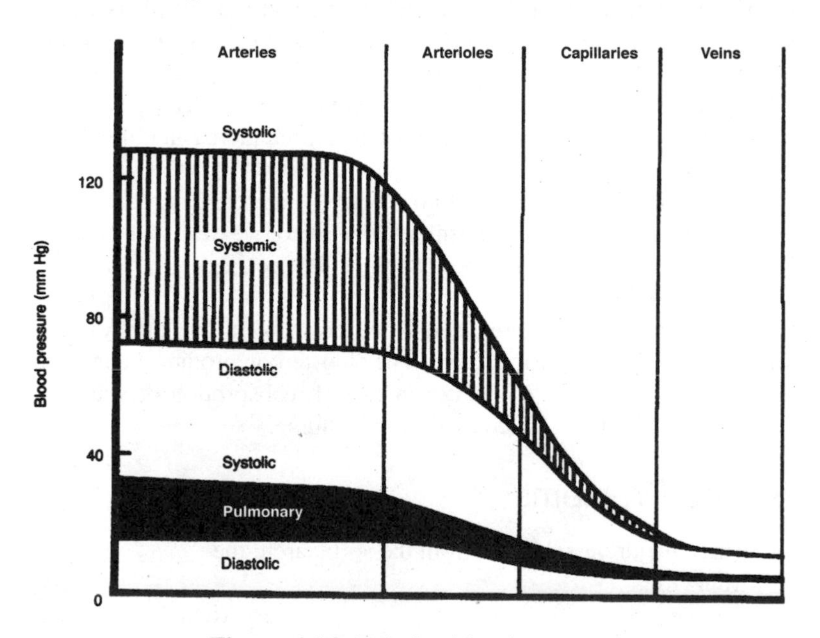


**Figure 16-3.** Relative blood pressures in systemic and pulmonary vessels.

Arterial blood pressure is much greater than venous blood pressure due to ventricular contractions of the heart pulsating the blood into arteries and the recoiling of the arterial walls. Systolic and diastolic blood pressures are greater in systemic arteries (approx. 120/80) than in pulmonary arteries (approx. 30/15). Blood pressure decreases in arteries proportionate to their distance from the heart. Blood pressure decreases rapidly within the capillaries and is near zero where venous blood enters the heart (Figure 16-3).

**Remember**

| | |
|---|---|
| **Systolic pressure** | **120 mmHg** |
| **Diastolic pressure** | **– 80 mmHg** |
| **Pulse pressure** | **40 mmHg** |

## Regulation of Blood Flow

**Neural mechanism.** *Baroreceptors* (sensory receptors that monitor blood pressure) in the walls of large vessels and the chambers of the heart detect a *decrease in blood pressure*. This initiates the following responses:

1. Increased secretion of ADH from the pituitary. Under the action of ADH, the kidneys restore water to the blood, increasing blood volume.
2. Sympathetic impulses sent to the heart, which increase heart rate.
3. Sympathetic response of smooth muscle of vessels that changes vessel diameter and thus peripheral resistance.

**Renal mechanism.** A decrease in blood pressure in the kidneys activates the renin angiotensin system. The aldosterone produced alters the electrolyte balance and stimulates increased reabsorption of water by the kidneys, resulting in increased blood volume.

# Solved Problems

1. Three major vessels arise from the aortic arch: the ____________ trunk, the ____________ ____________ ____________ artery, and the ____________ ____________ artery. (**brachiocephalic, left common carotid, left subclavian**)
2. Venous blood returning from the arm passes through the brachial vein to the ____________ vein, and then to the subclavian vein. (**axillary**)
3. Systolic pressure minus diastolic pressure is the ____________ pressure. (**pulse**)
4. The tunica ____________ is the outer connective tissue layer of blood vessels. (**externa**)

# Chapter 17
# LYMPHATIC SYSTEM AND BODY IMMUNITY

IN THIS CHAPTER:

- ✔ *Lymphatic Structures*
- ✔ *Nonspecific Defenses*
- ✔ *Antibody-Mediated Immunity*
- ✔ *Cell-Mediated Immunity*
- ✔ *Transfusion Rejection Reaction*
- ✔ *Solved Problems*

Functioning together with the cardiovascular system, the lymphatic system:

1. transports interstitial fluid, called lymph, in lymph vessels from the tissues back to the blood where it contributes to blood plasma;
2. assists in fat absorption in the small intestine; and
3. plays a key role in protecting the body from bacterial invasion via the blood.

# Lymphatic Structures

## Lymphatic Vessels

Interstitial fluid enters the lymphatic system through the walls of *lymph capillaries*, composed of simple squamous epithelium. Lymph is carried then into larger *lymphatic ducts*. The lymph ducts eventually empty into one of two principal vessels: the **right lymphatic duct,** which drains lymph from the upper right quadrant of the body, and the large **thoracic duct,** which drains lymph from the remainder of the body. These drain into the right and left subclavian vein, respectively. Lymph flows through lymphatic vessels through the contraction of skeletal muscle, intestinal peristalsis, and gravity. *Valves* in lymphatic vessels prevent backflow.

## Lymph Nodes

**Lymph nodes** are small oval bodies enclosed in fibrous capsules. They contain phagocytic *cortical tissue* adapted to filter lymph. *Afferent lymphatic vessels* carry lymph into the node and *efferent lymphatic vessels* carry the filtered lymph from the node. **Lymphocytes,** leukocytes responsible for body immunity, and **macrophages,** phagocytic cells, are found in lymph nodes. Lymph nodes occur in clusters or chains. Some principal groups are:

- **Popliteal** and **inguinal nodes** of the lower extremity
- **Lumbar nodes** of the pelvic region
- **Cubital** and **axillary nodes** of the upper extremity
- **Cervical nodes** of the neck
- **Mesenteric nodes** associated with the small intestine

## Lymphoid Organs

Lymphoid organs are:

- **Tonsils:** The three tonsils, (**pharyngeal** (adenoid), **palatine**, and **lingual**), are lymph organs of the pharyngeal region. They function to fight infections of the nose, ear, and throat region.

- **Spleen:** The spleen is located in the upper left portion of the abdominal cavity. It is not a vital organ in the adult but assists other organs in producing lymphocytes, filtering blood and destroying old erythrocytes. It also serves as a reservoir for erythrocytes.
- **Thymus:** The thymus is in the anterior thorax, deep to the manubrium of the sternum. It is much larger in a child than an adult. In a child it functions to change undifferentiated lymphocytes into T lymphocytes and is a reservoir for lymphocytes.

## Nonspecific Defense Mechanisms

Types of non-specific defense mechanisms include:

- **Mechanical barriers:** skin and mucous membranes
- **Chemical barriers:** enzymes, HCl in stomach, lysozymes
- **Interferon:** proteins that inhibit viral growth
- **Phagocytes:** neutrophils, monocytes, and macrophages
- **Species resistance**

## Antibody-Mediated Immunity

Specific immunity refers to resistance of the body to specific foreign agents (**antigens**). These include microorganisms, viruses, and their toxins, as well as foreign tissue and other substances. Antigens are large complex molecules (proteins, polysaccharides) found on cell walls or membranes of foreign substances.

In **antibody-mediated immunity,** an antigen stimulates the body to produce special proteins, **antibodies,** that destroy the particular antigen through an *antigen-antibody reaction*. Antibodies are *gamma globulins* composed of four interlinked polypeptide chains, two short (light) chains and two long (heavy) chains (Figure 17-1). All antibodies have *constant regions*, that are structurally similar, and *variable regions,* that are the location of antigen-binding sites. Small variations in the variable region make each antibody highly specific for one particular antigen. Antigen-antibody binding induces the production of more antibodies specific for that antigen.

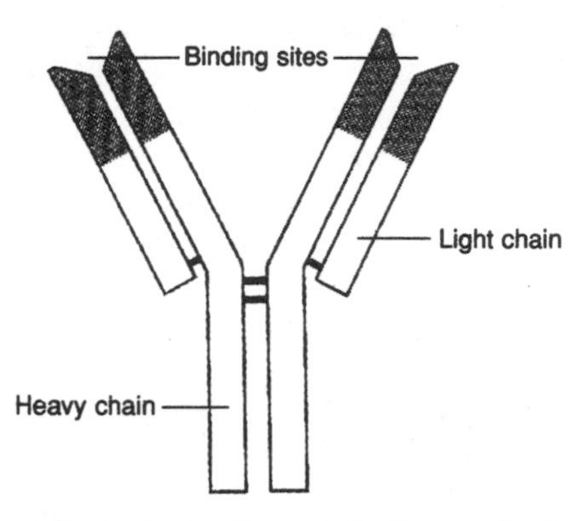


**Figure 1 -1.** A simple model of an antibody. Shaded areas indicate variable regions.

## You Need to Know 

**Active Immunity:** The body manufactures antibodies in response to direct contact with an antigen. When that antigen is encountered again, the body "remembers" it and responds more quickly. This is how vaccinations work.

**Passive Immunity:** Result of the transfer of antibodies from one person to another, as a result of an injection or transfer across the placenta.

## Cell-Mediated Immunity

**Cell-mediated immunity** is another mechanism of specific immunity. In this case, cells provide the main defensive strategy. Lymphocytes, circulating in the blood and found in lymphoid tissues, (T lymphocytes and B lymphocytes) become sensitized to an antigen, attach themselves to that antigen, and destroy it. **T lymphocytes** produce cell-mediated immunity. Upon interacting with a specific antigen they become sensitized and differentiated in several types of daughter cells.

**Table 17.1** Types and Functions of T Cells

| **Types of T cells** | **Function** |
|---|---|
| Memory T cells | Remain inactive until future exposure to same antigen. |
| Killer T cells | Combine with antigen, cause lysis of foreign cells, release cytokines. |
| Helper T cells | Help to activate other T or B cells. |
| Delayed-hypersensitivity T cells | Release cytokines. |

**B lymphocytes** produce antibody-mediated immunity. B lymphocytes become sensitized to an antigen, proliferate, and differentiate to form clones of daughter cells.

**Table 17.2** Types and Functions of B Cells

| **Types of B cells** | **Function** |
|---|---|
| Plasma cells | Produce antibodies specific to the antigen. |
| Memory B cells | Turn into plasma cells upon a later exposure to same antigen. |

## Other Components of the Immune System

Other components of the immune system are:

- **Cytokines** (*interferons, chemotactic factors, macrophage-activating factors, migration-inhibiting factors, transfer factors*): Chemical messengers used by the immune system to enhance immune response.
- **Complement system:** Enzyme precursors that aid immune response by causing lysis of invading cells, attract and enhance the action of phagocytes, enhance inflammation, and neutralize viruses.

# Transfusion Rejection Reaction

Red blood cells have large numbers of antigens present on their cell membranes; these can initiate antibody production, and therefore antigen-antibody reactions. One of the groups of antigens most likely to cause blood transfusion reactions is the **ABO system.** Antigens are inherited factors present on the RBC membranes at birth. If the recipient of a blood transfusion and the donor are improperly matched the antigen-antibody reaction (transfusion reaction) occurs (see Table 17.3) causing RBC to clump, rupture, and hinder the flow of blood.

**Table 17.3** The ABO Antigen System and Potential Donors

| Blood group | Antigens | Antibodies | Permissible donors |
|---|---|---|---|
| A | A | Anti-B | A or O |
| B | B | Anti-A | B or O |
| AB | A and B | none | A, B, O |
| O | none | Anti-A and Anti-B | O |

Another group of antigens associated with blood is the *Rh system*. Rh antigens are present on the red blood cell membranes of about 85% of the population. These people are classed as Rh positive (Rh+). The remaining 15% are classed as Rh negative (Rh-). Rh- individuals do not develop antibodies against Rh antigens until they are exposed to Rh+ blood.

# Solved Problems

**True or False**

____ 1. A person with type B blood has B antibodies. (**False**)

____ 2. A person who encounters a pathogen and who has a primary immune response develops passive immunity. (**False**)

____ 3. The interaction of antigen with antibody is highly specific. (**True**)

____ 4. Valves are present in lymphatic vessels. (**True**)

____ 5. Passive immunity is the transfer of antibodies developed in one individual into the body of another. (**True**)

____ 6. When antigenically stimulated, B lymphocytes proliferate and form plasma cells. (**True**)

____ 7. Antigens are small lipid molecules that stimulate the immune response. (**False**)

# Chapter 18
# Respiratory System

IN THIS CHAPTER:

- ✔ *Respiration*
- ✔ *Components of the Respiratory System*
- ✔ *Mechanics of Breathing*
- ✔ *Respiratory Volumes*
- ✔ *Gas Transport*
- ✔ *Regulation of Respiration*
- ✔ *Solved Problems*

## Respiration

All cells require a continuous supply of oxygen ($O_2$) and must continuously eliminate a metabolic waste product, carbon dioxide ($CO_2$). On the macroscopic level, the term **respiration** simply means ventilation, or breathing. On the cellular level, it refers to the processes by which cells utilize $O_2$, produce $CO_2$, and convert energy into useful forms.

**Remember**

| | |
|---|---|
| **External respiration:** | gas exchange between the blood and the air |
| **Internal respiration:** | gas exchange between the blood and the cells |
| **Cellular respiration:** | cell use of $O_2$ for metabolism, yielding $CO_2$ as a waste product |

## Components of the Respiratory System

The major passages of the respiratory system are the **nasal cavity**, **pharynx, larynx,** and **trachea.** Within the lungs, the trachea branches into **bronchi, bronchioles,** and finally **pulmonary alveoli.** While the primary function of the respiratory system is *exchange of gasses* for cellular metabolism, portions of the respiratory system also function in *sound production*, *abdominal compression*, and *coughing* and *sneezing*. The **conducting division** of the respiratory system includes all cavities and structures that transport gases to and from the pulmonary alveoli.

### Nasal Cavity

Structures: nasal superior septum; middle and inferior nasal conchae.
Tissues: pseudostratified ciliated columnar epithelium; olfactory epithelium
Warms and moistens the inspired air, also functions in olfaction.

## Pharynx

**Nasopharynx:** auditory (eustachian) canals, uvula, pharyngeal tonsils
**Oropharynx:** palatine and lingual tonsils
**Laryngeopharynx:** larynx
The oropharynx and laryngopharynx have respiratory and digestive functions, while the nasopharynx serves only the respiratory system.

## Larynx

Structures: anterior thyroid cartilage, epiglottis, cricoid cartilage, arytenoid cartilages, cuneiform and corniculate cartilages, and the glottis.
The larynx forms the entrance into the trachea. Its primary function is to prevent food or fluid from entering the trachea and lungs during swallowing. A secondary function is sound production.

## Trachea and Bronchial Tree

Structures: trachea branches into right and left primary bronchi, further branching into secondary bronchi, tertiary bronchi, and bronchioles.
Tissues: cartilaginous rings; lined with mucous-secreting pseudostratified ciliated columnar epithelium.
Serves as a conducting system for air. Cartilaginous rings hold passages open.

## Respiratory Division

Structures: continued branching into terminal bronchioles, alveolar ducts, alveolar sacs, pulmonary alveoli.
Tissues: Simple cuboidal epithelium in alveolar ducts, simple squamous epithelium in pulmonary alveoli.
Gas exchange occurs in the pulmonary alveoli, external respiration. **Septal cells,** which secrete a surfactant that lower the surface tension, and **alveolar macrophages** that remove foreign debris from the alveolus are found in the alveolar walls.

## Lungs

The paired **lungs** are contained within the thoracic cavity, separated from each other by the *mediastinum* (Figure 18-1). Each lung is composed of **lobes,** and these, in turn, of **lobules** that contain the alveoli. The left lung has a **cardiac notch** on its medial surface. It is subdivided into two lobes by a single **fissure** and contains eight **bronchial segments.** The right lung is subdivided into three lobes by two fissures, and contains ten bronchial segments.

The lungs are surrounded by a two-layered serous membrane, the **pleurae.** The inner layer, **visceral pleura,** is attached to the surface of the lungs; the outer layer, **parietal pleura,** lines the thoracic cavity. Between the visceral and parietal pleura is a moist potential space, the **pleural cavity.** Air pressure in the pleural cavity (intrathoracic pressure) is slightly lower than atmospheric pressure in resting lungs. This negative pressure is critical for the thoracic cavity to "pull out" on the lungs causing them to inflate.

## Mechanics of Breathing

**Inspiration** occurs when contraction of the inspiratory muscles causes an increase in thoracic volume, with expansion of the lungs and a decrease in intrathoracic and intrapulmonic (alveolar) pressures. Air enters the lungs when intrapulmonic pressure falls below atmospheric pressure (760 mmHg at sea level). **Expiration** follows passively, causing thoracic volume to decrease and intrapulmonic pressure to rise.

### You Need to Know 

| | |
|---|---|
| **Muscles of inspiration:** | Diaphragm and external intercostals. |
| **Expiration:** | Passive recoil of inspiratory muscles. |
| **Forced expiration:** | Internal intercostals & abdominal muscles. |

## Respiratory Volumes

*Total lung capacity* may be expressed as the sum of four volumes (Figure 18-1): **tidal volume,** the volume of air moved into and out of the lungs during normal breathing; **inspiratory reserve,** the maximum volume beyond the tidal volume that can be inspired in one breath; **expiratory reserve,** the maximum volume beyond tidal volume that can be forcefully exhaled following a normal expiration; and **residual volume,** the air that remains in the lungs following a forceful expiration. Respiratory air volumes are measured with the *spirometer*.

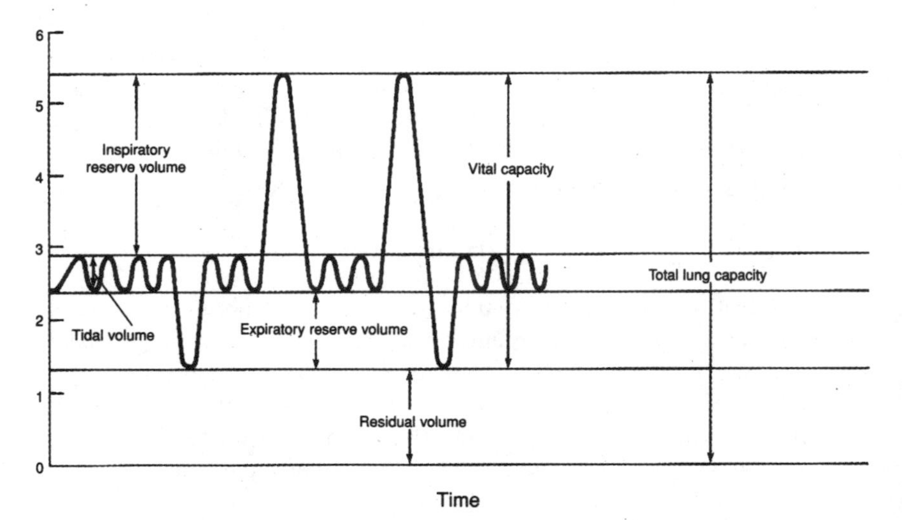


**Figure 18-1.** Respiratory air volumes.

Minute respiratory volume is the volume of air moved in one minute. A normal value is 6 L/sec. **Alveolar ventilation** is the volume of air exchanged in 1 minute in the pulmonary alveoli.

**Minute respiratory volume** = (tidal volume) × (respiratory rate)
**Alveolar ventilation** = [(tidal volume) – (dead space)] × (respiratory rate)

*Dead space* is the volume of air in the conducting division, typically 150 ml in an adult.

## Gas Transport

Up to 99 percent of $O_2$ in the blood is transported on hemoglobin molecules in erythrocytes. $CO_2$ in blood is mostly converted to bicarbonate ions in erythrocytes and released into the blood plasma.

**Partial Pressures**

In a mixture of gases, each component gas exerts a partial pressure that is proportional to its concentration in the mixture. For example, air is 21 percent $O_2$, this gas is responsible for 21 percent of air pressure. Twenty-one percent of 760 mmHg = 160 mmHg, the partial pressure of $O_2$. The difference in partial pressures in the alveolus and in the pulmonary capillaries favors diffusion of $O_2$ from the alveolus into the blood and $CO_2$ from the blood into the alveolus.

| Atmospheric | Alveolar | Pulmonary Capillary |
|---|---|---|
| $P_{O2}$ = 160 mmHg | $P_{O2}$ = 101 mmHg | $P_{O2}$ = 40 mmHg |
| $P_{CO2}$ = 0.3 mmHg | $P_{CO2}$ = 40 mmHg | $P_{CO2}$ = 45 mmHg |

**Acid-base balance**

The presence of the enzyme **carbonic anhydrase** in erythrocytes causes about 67 percent of the $CO_2$ in blood to combine quickly with water to form carbonic acid, most of which dissociates into bicarbonate and hydrogen ions:

$$CO_2 + H_2O \leftrightarrow H_2CO_3 \leftrightarrow HCO_3^- + H^+$$

Bicarbonate ions ($HCO_3^-$) make up a large part of the blood buffer system. **Respiratory acidosis** (blood pH below 7.35) occurs when $CO_2$ is not eliminated from the body at a normal rate, increasing vascular $P_{CO2}$. This can be caused by lung disease or hypoventilation. Respiratory alkalosis (blood pH above 7.45) occurs when $CO_2$ is eliminated too rapidly, decreasing vascular $P_{CO2}$. This may result from hyperventilation or certain drugs that affect the respiratory control center of the brain.

## Regulation of Respiration

Control of respiration occurs in the **expiratory** and **inspiratory centers** of the rhythmicity area of the medulla oblongata of the brain. When the inspiratory neurons are excited, the respiratory muscles are stimulated to produce inhalation and the expiratory neurons are inhibited. After about 2 seconds the reciprocal process occurs. The medulla also contains chemoreceptors concerned with respiration as do the **carotid**

**bodies** in the carotid arteries and **aortic bodies** in the arotic arch. These receptors respond to increased $P_{CO2}$ by initiating inspiration.

## Solved Problems

_____ 1. Total lung capacity (**a**)
_____ 2. Expiratory reserve volume (**d**)
_____ 3. Vital capacity (**e**)
_____ 4. Residual volume (**f**)
_____ 5. Tidal volume (**b**)
_____ 6. Inspiratory reserve volume (**c**)

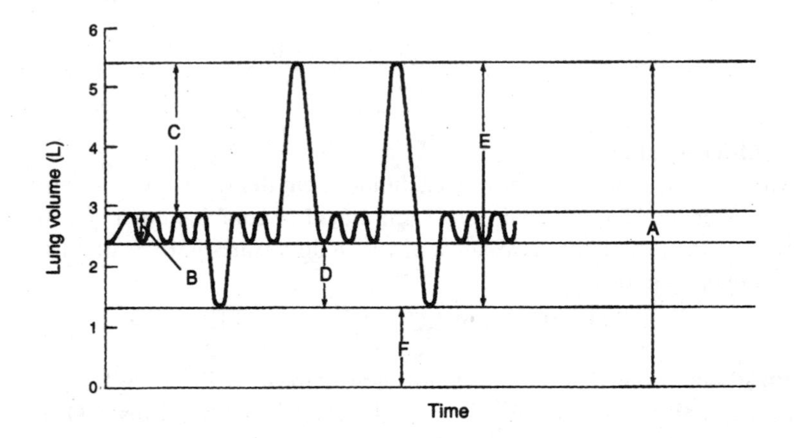

# Chapter 19
# DIGESTIVE SYSTEM

IN THIS CHAPTER:

- ✔ *Digestive Processes*
- ✔ *Peritoneum*
- ✔ *Histology of the GI Tract*
- ✔ *Structures and Functions of the GI Tract*
- ✔ *Accessory Organs*
- ✔ *Solved Problems*

## Digestive Processes

The food we eat is utilized at the cellular level in chemical reactions involving: synthesis of proteins, carbohydrates, hormones, and enzymes; cellular division, growth, and repair; and production of heat. To become usable by the cells, most food must first be mechanically and chemically reduced to forms that can be absorbed through the intestinal wall and transported to the cells by the blood. The processes involved with digestion include:

**Ingestion:** Taking food into the mouth (mechanical process)
**Mastication:** Chewing food (mechanical process)
Salivary action (chemical process)
**Deglutition:** Swallowing (mechanical process)

**Peristalsis:** Wavelike contractions that move food through the GI tract (mechanical process)

**Absorption:** Passage of food molecules from the GI tract into the circulatory or lymphatic system (mechanical and chemical process)

**Defecation:** Elimination of undigestible wastes (mechanical process)

The digestive system can be divided into a tubular gastrointestinal tract (GI tract) and accessory digestive organs. The GI tract extends from the oral cavity (mouth) to the anus. The regions or organs of the GI tract include the oral cavity, pharynx, esophagus, stomach, small intestine, and large intestine. The rectum and anal canal are located at the terminal end of the large intestine. The accessory digestive organs include the teeth, tongue, salivary glands, liver, gallbladder, and pancreas.

## Peritoneum

Associated with the abdominal cavity and its viscera are serous membranes (see Chapter One), called **peritoneum.** The parietal peritoneum lines the wall of the abdominal cavity. The peritoneal covering continues around the abdominal viscera as the visceral peritoneum. The space between the parietal and visceral portions of the peritoneum is called the **peritoneal cavity.** Most of the digestive organs are located within the peritoneal cavity. A few are outside of the peritoneum, retroperitoneal.

**Table 19.1** Extensions of the Parietal Peritoneum

| | |
|---|---|
| **Falciform ligament** | Attaches the liver to the diaphragm and anterior body wall |
| **Lesser omentum** | Extends between the liver and the lesser curvature of the stomach |
| **Greater omentum** | Extends from the greater curvature of the stomach to the transverse colon |
| **Mesentery** | Supports the intestine |
| **Mesocolon** | Supports the large intestine |

## Histology of the GI Tract

**Table 19.2** Four Tunics (Histological Layers)

From the innermost to the outermost:

| **Tunic** | **Structure** | **Function** |
|---|---|---|
| **Mucosa** | Simple columnar epithelium | Secretion and absorption |
| **Submucosa** | Highly vascular; autonomically innervated | Absorption |
| **Muscularis** | Smooth muscle | Peristalsis |
| **Adventitia (visceral serosa)** | Visceral peritoneum | Binding and protection |

## Structures and Functions of the GI Tract

**Oral Cavity.** The oral cavity ingests food; grinds and mixes it with saliva; initiates digestion of carbohydrates; forms bolus; swallows bolus.

- **Teeth:** Four types of teeth: **Incisors** (4 upper, 4 lower) for cutting and shearing food; **Canines** (2 upper, 2 lower) for holding and tearing; **Premolars** (4 upper, 4 lower) and **Molars** (6 upper, 6 lower) both for crushing and grinding food.
- **Tongue:** Moves food around mouth during mastication; assists in swallowing; formation of speech sounds; location of taste buds.
- **Salivary glands:** Three salivary glands: **Parotid gland** located over the masseter muscle; **Submandibular gland** inferior to the base of the tongue; **Sublingual gland** under the tongue, produces saliva and begins chemical digestion of carbohydrates.
- **Palate:** Roof of the oral cavity consisting of the bony **hard palate** anteriorly and the **soft palate** posteriorly. The **uvula** is suspended from the soft palate. The soft palate closes the nasopharynx during swallowing.

**Pharynx:** A funnel-shaped passageway that connects the oral and nasal cavity to the esophagus and trachea. Function is swallowing.

**Esophagus:** A muscular tube located in the thorax behind the trachea that connects the pharynx to the stomach. Transports bolus to the stomach via peristalsis.

**Stomach:** See Figure 19-1 for parts of the stomach. The stomach receives the bolus from esophagus; churns bolus with gastric juice to form chyme; initiates digestion of proteins; carries out limited absorption; and moves chyme into duodenum. Specializations of the tunics of the stomach include: an additional smooth muscle layer, the **oblique layer;** longitundinal folds of the mucosa called **rugae** and **gastric glands,** which secrete gastric juice.

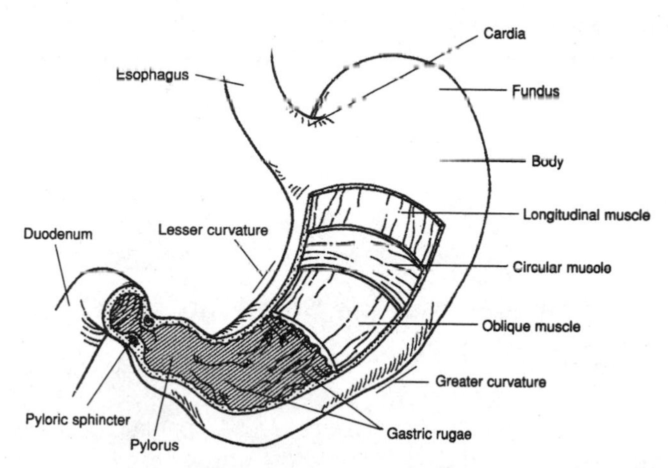


**Figure 19-1.** The stomach.

**Table 19.3** Composition of Gastric Juice

| Component | Source | Function |
|---|---|---|
| **Hydrochloric acid (HCl)** | Parietal cells | Converts pepsinogen to pepsin; kills pathogens |
| **Pepsinogen** | Chief cells | Inactive form of pepsin |
| **Pepsin** | From pepsinogen in the presence of HCl | Protein-splitting enzyme |
| **Mucus** | Goblet cells | Protects the mucosa |
| **Intrinsic factor** | Parietal cells | Aids absorption of vitamin B12 |
| **Serotonin & histamine** | Argentaffin cells | Autocrine regulators |
| **Gastrin** | G cells | Stimulates secretion of HCl and pepsin |

**Small intestine**: The region between the stomach and the large intestine, approximately 10 feet long. It receives chyme from the stomach, bile and pancreatic secretions from the liver and pancreas respectively; chemically and mechanically breaks down chyme; absorbs nutrients; and transports wastes to the large intestine.

**Divisions of the Small Intestine**

**Duodenum:** From **pyloric sphincter** of stomach to duodenojejunal flexure (approx. 10 in.). *Common bile duct* and *pancreatic duct* empty into duodenum.

**Jejunum:** Middle section of the small intestine. Characterized by deep folds of the mucosa and submucosa, **plicae circulares.**

**Ileum:** Joins the **cecum** of the large intestine at the **ileocecal valve.**

Modifications of the small intestine facilitate absorption. The **plicae circulares** and the **villi,** fingerlike projections of the mucosa, increase absorptive area. Each villus contains a capillary network and a lymph vessel, a **lacteal.** Absorption is accomplished as food molecules enter these vessels through **microvilli,** microprojections on the surface of the villi. At the bases of the villi are **intestinal glands** that secrete digestive enzymes.

## You Need to Know 

### Intestinal Enzymes and Their Actions

| | |
|---|---|
| **Peptidase:** | Converts proteins into amino acids |
| **Sucrase (maltase and lactase):** | Converts disaccharides into monosaccharides |
| **Lipase:** | Converts fats into fatty acids and glycerol |
| **Amylase:** | Converts starch and glycogen into disaccarides |
| **Nuclease:** | Converts nucleic acids into nucleotides |
| **Enterokinase:** | Activates trypsin secreted from the pancreas |

**Large intestine:** Extends from the ileocecal valve to the anus, approx. 5 feet long. It receives wastes from the small intestine; absorbs water and electrolytes; forms, stores, and expels feces through defecation.

### Divisions of the Large Intestine

**Cecum:** First portion, resembles a dilated pouch.
**Ascending colon:** Extends superiorly on the right from the cecum to the liver, at the hepatic flexure.
**Transverse colon:** Transversely crosses the upper peritoneal cavity.
**Descending colon:** Extends inferiorly on the left from the splenic flexure to the pelvis.
**Sigmoid colon:** S-shaped bend in the pelvic region
**Rectum:** Terminal portion of the large intestine.

The **appendix,** consisting of lympatic tissue, is attached to the cecum. The tunics, mucosa, and submucosa have sacculations called **haustra;** the muscularis consists of longitunal bands called **taeniae coli,** and attached to the adventitia are fat-filled pouches called **epiploic appendages.**

## Accessory Organs

**Liver.** Lobes of the liver are:

- **Lobes of the liver:**
- **Right** and **left lobe,** separated by the **falciform ligament.**
- **Caudate lobe** is near the inferior vena cava.
- **Quadrate lobe** is between the left lobe and the gall bladder.

The liver receives oxygenated blood from the hepatic artery, a branch of the celiac artery. It also receives food-laden blood from the hepatic portal vein, which carries venous blood from the GI tract. In the **liver sinusoids,** blood from both sources mix. Oxygen, nutrients, and certain toxic substances are extracted by **hepatic cells.**

Functions of the liver are:

- Synthesis, storage, and release of vitamins and glycogen
- Synthesis of blood proteins
- Phagocytosis of worn red and white blood cells and bacteria
- Removal of toxic compounds
- Production of bile, which emulsifies fats in the duodenum

**Gall bladder.** Pouchlike organ attached to the inferior surface of the liver. Stores and concentrates bile. Bile, produced in the liver, drains through the **hepatic ducts** and **bile duct** to the duodenum. When the small intestine is empty, bile is forced up the **cystic duct** to the gallbladder for storage.

**Pancreas.** Lies horizontally along the posterior abdominal wall, adjacent to the greater curvature of the stomach. Endocrine function is discussed in Chapter Thirteen. Exocrine secretions act on all three food groups.

## Solved Problems

1. A __________ contains a lacteal for the absorption of fats and lymph into the lymphatic system. (**villus**)

2. The ______________ is the serous membrane that lines the wall of the abdominal cavity and covers visceral organs. (**peritoneum**)
3. The ______________ is the specific part of the mesentery that supports the large intestine. (**mesocolon**)
4. Food and fluid in the stomach are consolidated into a pasty material called ______________. (**chyme**)
5. Capillaries within the ______________ of the small intestine are sites where nutrients and fluids are absorbed into the circulatory system. (**villi**)

# Chapter 20
# Metabolism and Temperature Regulation

In This Chapter:

- ✔ *Metabolism*
- ✔ *Carbohydrate Metabolism*
- ✔ *Lipid Metabolism*
- ✔ *Protein Metabolism*
- ✔ *Hormonal Regulation of Metabolism*
- ✔ *Temperature Regulation*
- ✔ *Solved Problems*

## Metabolism

Foods are first digested, then absorbed, and finally, metabolized. **Metabolism** refers to all the chemical reactions of the body. There are two aspects of metabolism: *catabolism*, the breaking-down process which provides energy (stored in ATP), and *anabolism*, the building-up process which requires energy. All metabolic reactions within the body, whether anabolic or catabolic, are catalyzed by enzymes. The substances in food that enter into metabolism are referred to as nutrients. They are classified as carbohydrates, lipids (fats), proteins, vitamins, minerals, and water.

## Carbohydrate Metabolism

The average human diet consists largely of polysaccharide and disaccharide carbohydrates. When digested, these molecules are broken down into the monosaccharides glucose, fructose, and galactose. The liver further converts fructose and galactose into glucose. Glucose is the molecule from which energy is formed. The equations for glucose metabolism are:

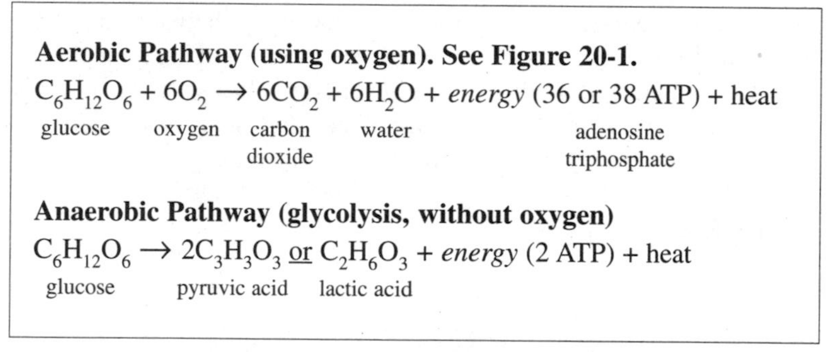

**Aerobic Pathway (using oxygen). See Figure 20-1.**

$C_6H_{12}O_6 + 6O_2 \rightarrow 6CO_2 + 6H_2O +$ *energy* (36 or 38 ATP) + heat

glucose, oxygen, carbon dioxide, water, adenosine triphosphate

**Anaerobic Pathway (glycolysis, without oxygen)**

$C_6H_{12}O_6 \rightarrow 2C_3H_3O_3$ or $C_2H_6O_3$ + *energy* (2 ATP) + heat

glucose, pyruvic acid, lactic acid

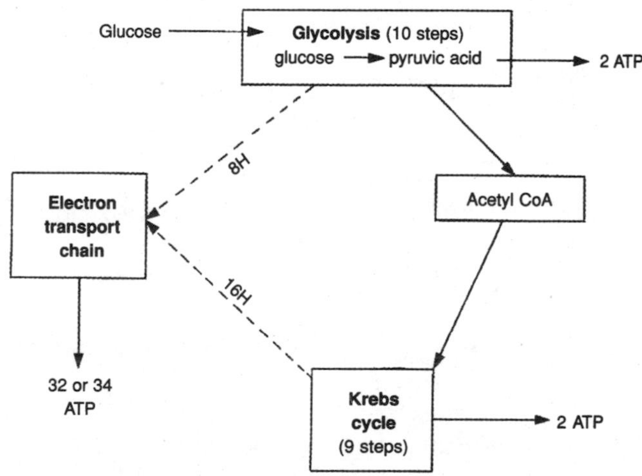


**Figure 20-1.** Aerobic production of ATP.

Glycolysis is more rapid than the aerobic pathway, but supplies much less energy and produces lactic acid, which causes early fatigue. Anaerobic activity can only be performed for a short period of time. The 10 steps of glycolysis take place in the cytoplasm of the cell. The Krebs cycle (9 steps, 8 enzymes) and the electron transport chain of oxidation-reduction reactions both take place in the mitochondria.

Not all glucose entering the cell is immediately catabolized to form energy. Extra glucose is linked together into a storage molecule, *glycogen*. When the body needs energy, glycogen, stored in the liver and muscle cells, is broken down and glucose is released into the blood. This inverse process, *glycogenolysis*, is spurred by the pancreatic hormone glucagon, and adrenal hormones epinephrine and norepinephrine.

Both protein and lipids can be converted to glucose by the process *gluconeogenesis*. Five hormones stimulate gluconeogenesis: cortisol, thyroxine, glucagon, growth hormone, and epinephrine.

## Lipid Metabolism

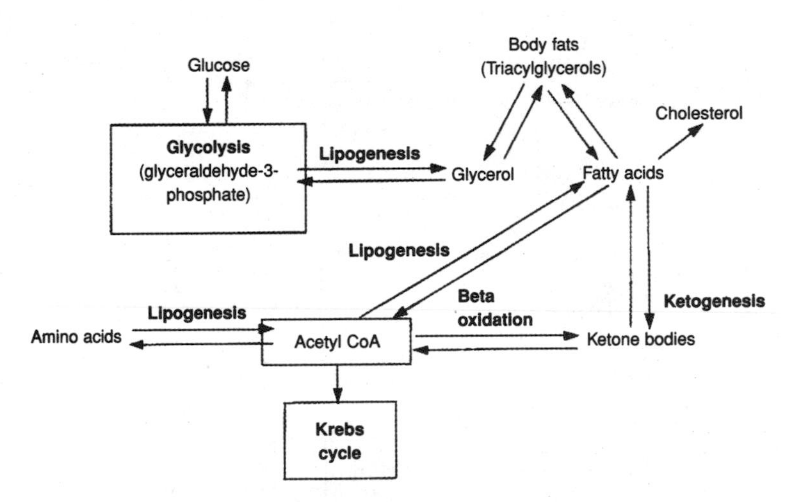


**Figure 20-2.** Lipid metabolism.

Lipids are second to carbohydrates as a source of energy for ATP synthesis. Fats participate in the building of many cellular structures and hormones. Lipid metabolism is diagramed in Figure 20-2.

When total food intake exceeds the body's needs, the excess is converted into fat and stored. When stored fats are catabolized, the glycerol components may enter the glycolytic pathway to produce energy or glucose, the fatty acid components break down to form acetyl CoA. This is called *beta oxidation*. The anabolic process, leading from glucose or amino acids to lipids, is called *lipogenesis*.

## Protein Metabolism

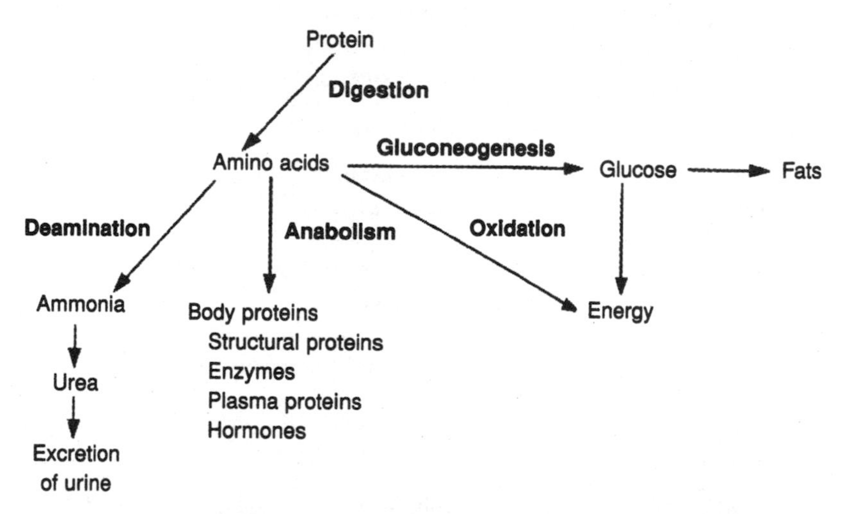


**Figure 20-3.** Protein metabolism.

Proteins play an essential role in cellular structure and function. Protein metabolism is diagrammed in Figure 20-3. Amino acids in proteins may be used as an energy source when other sources prove inadequate through the Krebs cycle.

## Hormonal Regulation of Metabolism

**Table 20.1** Hormone Regulation of Metabolism

| **Hormone** | **Metabolic effects** |
|---|---|
| **Insulin** | Promotes glucose uptake into cells; glycogenesis, lipogenesis, amino-acid uptake into cells and protein synthesis. Inhibits lipolysis. |
| **Glucagon and Epinephrine** | Promotes glycogenolysis, gluconeogenesis, and protein synthesis. |
| **Thyroxine** | Promotes glycogenolysis, gluconeogenesis, and lipolysis. |
| **Growth hormone** | Promotes amino-acid uptake into cells, protein synthesis, glycogenolysis, lipolysis. |
| **Cortisol** | Promotes gluconeogenesis, lipolysis, and breakdown of proteins. |
| **Testosterone** | Promotes protein synthesis. |

Other factors that affect metabolic rate include increases in body size, body temperature, activity, and sympathetic stimulation.

## Temperature Regulation

Heat is continually being produced as a by-product of metabolism and is continually being lost to the surroundings. The body normally balances heat gain and loss.

## Solved Problems

### True or False

_____ 1. The hormone glucagon facilitates glucose uptake by the cells. (**False**)

_____ 2. Both lipids and proteins can be converted into glucose. (**True**)

_____ 3. Aerobic metabolism produces more ATP than does anaerobic metabolism. (**True**)

_____ 4. All carbohydrates ingested by the body are converted into and catabolized as glucose. (**True**)

_____ 5. All processes that deal with the catabolism of glucose (glycolysis, Krebs cycle) take place in the mitochondria of the cell. (**False**)

# Chapter 21
# URINARY SYSTEM

IN THIS CHAPTER:

- ✔ *Components of the Urinary System*
- ✔ *The Nephron and Its Function*
- ✔ *Urine Concentration*
- ✔ *Acid-Base Balance*
- ✔ *Micturition*
- ✔ *Solved Problems*

The urinary system plays a critical role in regulating the composition of body fluids (water balance, electrolyte balance, and acid-base balance). It also rids the body of metabolic wastes and foreign matter (chemicals, drugs). The kidneys also have a minor endocrine function.

## Components of the Urinary System

**Kidneys.** The kidneys are located on either side of the vertebral column in the abdominal cavity, between the twelfth thoracic and third lumbar vertebrae. They form urine. The macroscopic structure of the kidneys is diagramed and labeled in Figure 21-1.

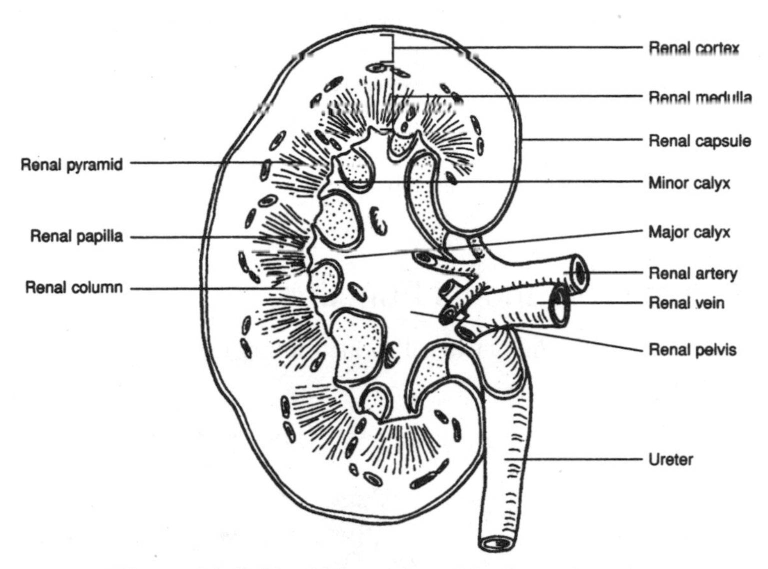


**Figure 21-1.** The kidney viewed in coronal section.

**Ureters.** The ureters transfer urine from the renal pelvises of the kidneys to the urinary bladder. The ureters are retroperitoneal.

**Urinary bladder.** The urinary bladder lies posterior to the symphysis pubis and anterior to the rectum. It stores urine. The urinary bladder consists of four tunics: the (innermost) *mucosa,* which is folded into rugae that allow distension of the bladder; the *submucosa,* which provides a rich vascular supply; the muscularis, a *smooth muscle layer*, (the **detrusor muscle);** and the *serosa*, a continuation of the peritoneum. The floor of the urinary bladder is a triangular area called the **trigone**. It has an opening at each of its three angles for the two ureters laterally and the urethra at the apex. Sympathetic fibers innervate the trigone, urethral openings, and blood vessels; parasympathetic fibers innervate the smooth muscle wall.

**Urethra.** The urethra conveys urine from the urinary bladder to the outside of the body. The *internal urethral sphincter*, composed of smooth muscle, and the *external urethral sphincter*, of skeletal muscle, constrict the lumen of the urethra causing the urinary bladder to fill. The urethra of a female is about 4 cm long, and that of a male about 20 cm. long. In a male, the spongy urethra also carries semen during ejaculation.

**Table 21.1** Blood Flow through the Kidney

| Renal artery → Interlobar arteries → Arcuate arteries → Interlobular arteries → Afferent arteriole → Glomerulus → Efferent arterioles → Peritubular capillaries and vasa recta → Interlobular veins → Arcuate veins → Interlobar veins → Renal vein |
| --- |

## The Nephron and Its Function

Figure 21-2 depicts the functional (urine-forming) unit of the kidney, the **nephron**.

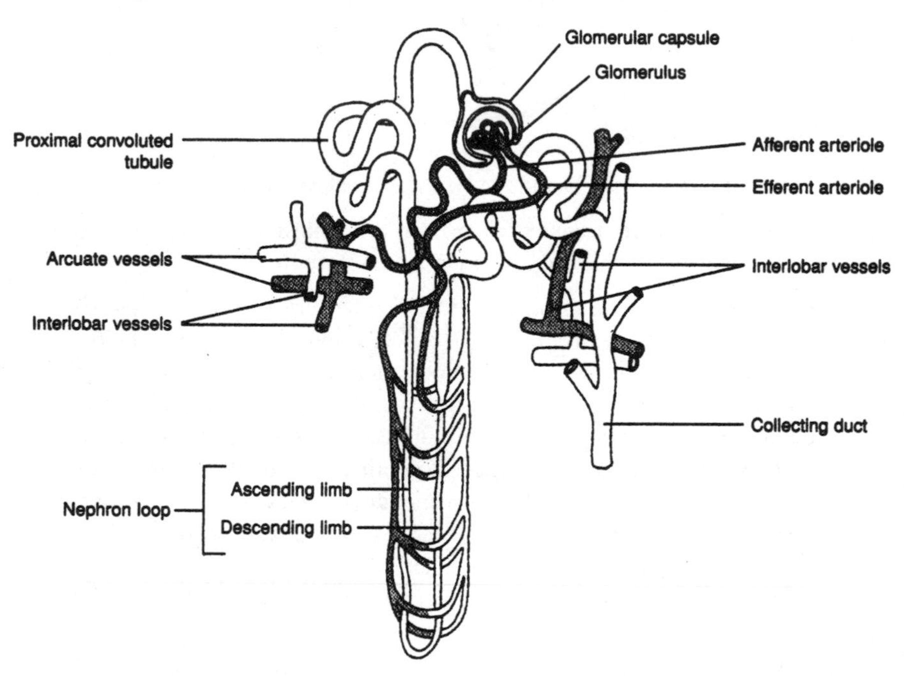


**Figure 21-2.** The nephron.

There are over 1 million nephrons per kidney. The components of the nephron are described below.

**Table 21.2** Components of the Nephron

| | |
|---|---|
| **Glomerulus** | Capillary network, highly permeable. |
| **Glomerular capsule** | Double-walled cup-like structure composed of squamous epithelium. Inner layer composed of modified cells, **podocytes,** that are closely associated with the glomerular capillaries. **Site of glomerular filtration.** |
| **Proximal convoluted tubule** | Simple cuboidal epithelium containing microvilli to increase the surface area. **Primary site of tubular reabsorption and secretion.** |
| **Nephron loop** | Descending and ascending portions. Involved with the **urine concentrating mechanism.** |
| **Distal convoluted tubule** | Shorter than the proximal convoluted tubule. Contains the *macula densa*, specialized sensory cells which monitor NaCl concentration**. Site of some tubular reabsorption and secretion.** Empties into the collecting duct, which drains into the renal pyramid. |

**Juxtaglomerular apparatus:** Cells of the macula densa along with specialized juxtaglomerular cells of the afferent arteriole compose a sensory apparatus for monitoring blood pressure. A drop in blood pressure or an increased NaCl concentration in the distal tubule stimulates renin to be released from the juxtaglomerular cells. This activates the renin-angiotensin system (see Figure 13-2).

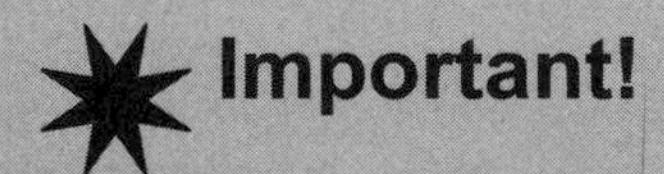


Three Components of Nephron Function:

- Glomerular filtration
- Tubular reabsorption
- Tubular secretion

**Glomerular filtration:** Fluid and solutes in the blood plasma of the glomerulus pass into the glomerular capsule. Filtrate has the same composition as blood plasma excluding proteins. **Glomerular filtration rate** (**GFR**) is the volume of filtrated formed by all nephrons each minute.

**Tubular reabsorption:** Approximately 99 percent of the filtrate is transported passively or actively out of the tubule into the interstitial fluid, then into the peritubular capillaries; 1 percent is excreted as urine. Most of the solutes are reabsorbed: 100 percent of the glucose, 99.5 percent of the sodium, and 50 percent of the urea.

**Tubular secretion:** Noxious substance such as hydrogen, potassium, poisons, drugs, and metabolic toxins are actively transferred from the peritubular capillaries into the interstitial fluid, then into the tubular lumen.

## Urine Concentration

The kidneys produce either a concentrated or dilute urine depending on the operation of a **countercurrent exchange mechanism** in the nephron loop, and the amount of circulating **antidiuretic hormone** (**ADH**) secreted from the posterior pituitary.

1. A concentration gradient exits in the renal medulla due to active transport of Cl- out of the tubular fluid in the ascending limb of the nephron loop, and the movement of Na+ ions out of the tubule. The ascending limb is impermeable to water and as Na+ and Cl- move out, the fluid in the ascending limb becomes more dilute.
2. Na+ and Cl- diffuse into the descending limb. The descending limb is permeable to water, and as water diffuses out into the interstitium as a result of the osmotic gradient, the tubular fluid in the descending limb becomes more concentrated.

3. Ions are actively transported into the intersititium from the collecting duct, urea passively diffuses out of the collecting duct into the interstitium.
4. Thin-wall vessels, the *vasa recta*, parallel the course of the nephron loops. Na+, Cl-, and water diffuse into the descending vasa recta and Na+ and Cl- diffuse out of the ascending vasa recta. These vessels function as countercurrent exchangers.

5. The amount of water reabsorbed form the distal convoluted tubules and collecting ducts is dependent on levels of ADH present. When low levels of ADH are secreted from the posterior, these tubules are impermeable to water, and a dilute urine is excreted. When ADH levels are high, these tubules are highly permeable to water, which is forced by the osmotic gradient out into the interstitium and a more concentrated urine is excreted.

## Acid-Base Balance

In acidosis, increased amounts of H+ are secreted into the kidney tubule and bicarbonate ions in the tubule are reabsorbed. In alkalosis, decreased amounts of H+ are secreted and less bicarbonate is reabsorbed. Two buffer systems in the tubular fluid carry excess H+ into the urine:

**Phosphate buffer system:** $HPO_32^- + H^+ \rightarrow H_2PO_4^-$

**Ammonia buffer system:** $NH_3 + H^+ \rightarrow NH_4^+$

## Micturition

Micturition is the physiological process of urination. Distention of the urinary bladder sends signals via sensory neurons to the spinal cord and up to the brain. Parasympathetic impulses stimulate the detrusor muscle to contract and the internal urethral sphincter to relax. If the decision is to urinate, the external urethral sphincter is relaxed and micturition results.

## Solved Problems

### Matching

| | |
|---|---|
| ____ 1. renin (**b**) | (a) adjacent to the macula densa of the capillaries |
| ____ 2. juxtaglomerular cells (**a**) | (b) secreted by juxtaglomerular cells |
| ____ 3. micturition (**e**) | (c) tube extending from kidney to urinary bladder |
| ____ 4. ureter (**c**) | (d) functional unit of the kidney |
| ____ 5. nephron (**d**) | (e) physiological events that result in the voiding of urine |

### Completion

1. The ____________ muscle within the wall of the urinary bladder forcefully contracts during micturition, forcing urine out of the urinary bladder. (**detrusor**)
2. ____________ is the hormone that regulates water reabsorption in the distal convoluted tubule. (**Antidiuretic hormone or ADH**)
3. The ____________ runs parallel to the nephron loops and functions as counter current exchangers. (**vasa recta**)

# Chapter 22
# Water and Electrolyte Balance

In This Chapter:

- ✔ *Distribution of Water in the Body*
- ✔ *Solute Concentrations*
- ✔ *Fluid Balance*
- ✔ *Electrolytes*
- ✔ *Solved Problems*

## Distribution of Water in the Body

Water is the most abundant substance in the human body, ranging from 40 percent–80 percent of total body weight (BW). All metabolic reactions require water. Body water is distributed between two major compartments: the intracellular fluid compartment (within the cells, 35 percent–40 percent BW) and the extracellular fluid compartment (outside the cells, 15 percent–20 percent BW). Extracellular fluid is distributed throughout the body.

**Distribution of Extracellular Fluid:**

| | | |
|---|---|---|
| Blood plasma | 4% – 5% BW | |
| Interstitial fluid | 11% – 15% BW | |
| Lymph | | |
| Transcellular fluid: | | Cerebrospinal fluid |
| | | Intraocular fluid |
| | | Synovial fliud |
| | | Pericardial, pleural, and peritoneal fluid |

Water functions in regulating body temperature, participating in hydrolysis reations, lubricating organs, providing cellular turgidity, and maintaining body homeostasis.

## Solute Concentrations

Two measures are used to describe solute concentration:

**Percent solution** = (grams of solute) / (100 ml of solution)
= (grams of solute / (dl of solution)

**Molarity** (**molar concentration**): MW is the molecular weight of the solute. One mole of solute weighs MW grams, from which

Moles solute = (grams of solute) / MW
Molarity (M) = (moles solute) / (liters of solution)

Fluid balance in the extracellular compartment is maintained by regulation of fluid osmolarity. Osmolarity of a body fluid is a measure of the concentration of individual solute particles dissolved in it. The osmolarities of extracellular and intracellular fluids are normally the same.

**Table 22.1** Mean Concentrations of Important Body Fluid Solutes

| Fluid | $Na^+$ | $K^+$ | $Ca^{2+}$ | $Mg^{2+}$ | $Cl^-$ | Amino Acids | Glucose mg% |
|---|---|---|---|---|---|---|---|
| Extracellular | 142 | 4 | 5 | 3 | 103 | 5 | 90 |
| Intracellular | 10 | 140 | 1 | 58 | 4 | 40 | 0–20 |

## Fluid Balance

Under normal conditions, fluid intake equals fluid output, so that the body maintains a constant volume. When water intake is greater than water output, a positive balance exists (hydration). Conversely, when output exceeds intake, a negative balance exists (dehydration). The amount of water consumed and the amount of urine formed are the two major mechanisms by which body water content is regulated.

Water is unconsciously regulated through the action of osmoreceptors located in the hypothalamus. These receptors sense the osmolality of the blood and determine whether more or less water is needed to maintain the correct osmolality. If blood is too concentrated, thirst is stimulated and we drink. ADH is released from the posterior pituitiary, which leads to conservation of body fluid in the collecting ducts of the kidneys, and decreased urine output. If blood is too dilute, thirst is suppressed and ADH release is inhibited, causing large volumes of dilute urine to be excreted.

A person may eliminate up to a liter of water over a 24-hour period without being aware of the loss. This is termed **insensible loss.** The loss occurs from the lungs and nonsweating skin.

When there is a loss of free water in the extracellular compartment, the fluid becomes too concentrated (increased osmolarity) and is referred to as being **hypertonic.** When there is a gain in free water, the fluid becomes too dilute and is referred to as being **hypotonic.**

## Electrolytes

**Electrolytes** are chemicals formed by ionic bonding that dissociate into electrically charged ions (cations and anions) when they dissolve in the body fluids. Examples of electrolytes are acids, bases, and salts. Nonelectrolytes are formed by covalent bonding. Most organic molecules are nonelectrolytes.

**Remember**

**Functions of Electrolytes:**

- Control osmolarity
- Maintain acid-base balance
- Metabolize essential minerals
- Particpate in all cellular activities.

## Solved Problems

### Completion

1. Body fluid that is too concentrated is referred to as __________. (**hypertonic**)
2. Water is autonomically regulated by __________ located in the hypothalamus of the brain. (**osmoreceptors**)
3. __________ become charged ions when they dissolve in the body fluids. (**Ions**)
4. __________ are formed by covalent bonding and do not ionize in the body fluids. (**Nonelectrolytes**)

### True or False

____ 1. Through insensible fluid loss, a person may eliminate up to a liter of water in a 24-hour period and not be aware of it. (**True**)

____ 2. The amount of water consumed and the amount of urine produced are the two major mechanisms by which body water is regulated. (**True**)

# Chapter 23
# Reproductive System

In This Chapter:

- ✔ *Gamete Formation*
- ✔ *Primary and Secondary Sex Organs*
- ✔ *Male Reproductive System*
- ✔ *Female Reproductive System*
- ✔ *Female Hormonal Cycle*
- ✔ *Fertilization and Pregnancy*
- ✔ *Solved Problems*

## Gamete Formation

**Gametes,** or *sex cells*, are the functional reproductive cells. They are *haploid cells*, each containing a half-complement of genetic material (23 single chromosomes). Fertilization of an **ovum** by a **spermatozoon** produces a normal *diploid cell*, the **zygote,** with 23 paired chromosomes. One out of the 23 pairs of human chromosomes determines sex. Sex chromosomes are of two types, X and Y.

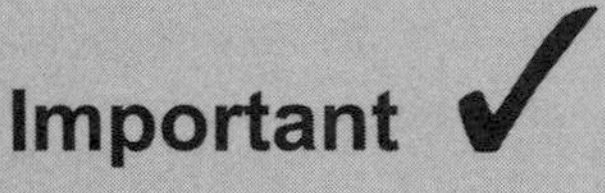

**Sex Determination:**

**XX = female:** All ova produced contain a single X chromosome.

**XY = male:** Equal number of X and Y spermatozoa are produced.

Sex of offspring determined by whether the fertilizing spermatozoa is X-bearing or Y-bearing.

**Spermatogenesis** is the process by which sperm cells are produced in the testes of a male. **Oogenesis** is the process by which ova are produced in the ovaries of a female. Both processes involve a special kind of cell division called **meiosis.** In meiosis (see Figure 3-1), each chromosome duplicates itself as in mitosis. However, in meiosis, the homologous chromosomes are attached to each other and come to lie alongside one another in pairs, producing a *tetrad* of four *chromatids*. Two *maturation divisions* are required to effect the separation of the tetrad into four daughter cells, each with one-half the original number of chromosomes. The nuclear aspects of meiosis are similar in males and females but there are differences in the cytoplasmic aspects such that:

Primary Spermatocyte → 4 spermatozoa
Primary Ooctye → 1 mature ovum

## Primary and Secondary Sex Organs

The **primary sex organs,** or **gonads,** are the *testes* in the male and the *ovaries* in the female. The gonads function as mixed glands, producing both hormones and gametes. The **secondary,** or *accessory*, **sex organs** are those structures that mature at puberty under the influence of sex hormones and that are essential in caring for and transporting gametes. **Secondary sex characteristics** are features that are considered sexual attractants.

## Male Reproductive System

The male sex organs are formed prenatally under the influence of testosterone secreted by the gonads (testes). During puberty, the secondary sex organs mature and become functional. Male reproductive organs (Figure 23-1) and their functions are listed below.

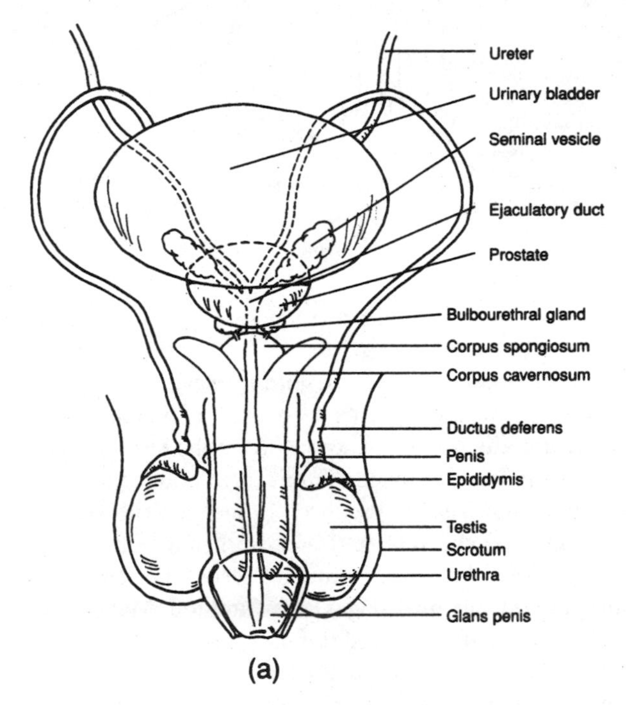


**Figure 23-1.** The male reproductive system. (a) anterior view

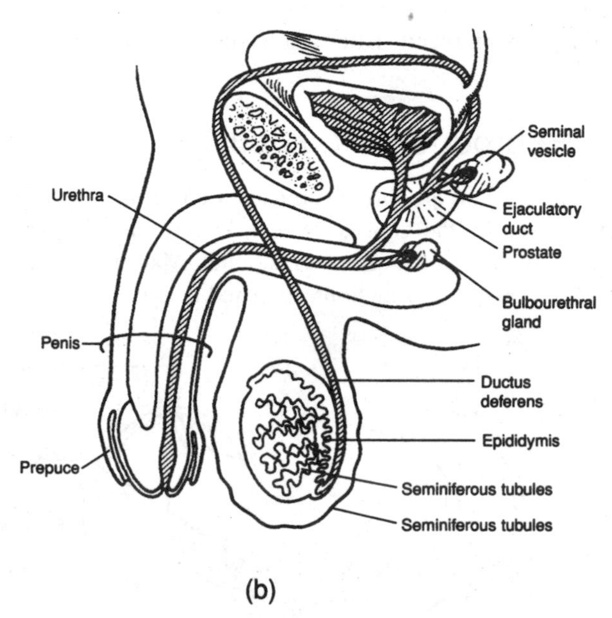


**Figure 23-1.** The male reproductive system. (b) sagittal view

**Testes.** The testes are located within the **scrotum,** a pouch of skin that encloses the testes. Each testis is covered by two tissue layers, the outer **tunica vaginalis,** a thin sac derived from the peritoneum; and the inner **tunica albuginea,** a tough fibrous membrane that encapsulates the testes. **Spermatozoa** are produced in the **seminiferous tubules** of the testes and enter the **rete testis** for further maturation. They are then transported out of the testes through a series of efferent ductules into the **epididymis** for the final stages of maturation. Mature spermatozoa are stored in the epididymis and the first portion of the **ductus deferens.** The testes also produce the *andogens*, male sex hormones, in the **interstitial cells. Sustentacular cells** of the testes provide essential molecules to the developing sex cells.

**Spermatic Ducts and Accessory Glands.** Once spermatozoa are mature, they pass through a series of tubules during ejaculation. From the ductus deferens, spermatozoa enter the **ejaculatory duct** where secretions from the seminal vesicle are added. The ejaculatory duct empties into the **prostatic urethra**. The prostate secretions also empty into the urethra. Spermatozoa then pass through the **membraneous urethra** into the **spongy urethra** in the penis. The bulbourethral glands are at the base of the penis. The accessory glands contribute alkaline secretions that form **semen** and function to nourish and enhance motility of spermatozoa and neutralize the acidic environment of the urethra and the vagina.

**Remember**

**Accessory Glands**

- seminal vesicles
- prostate
- bulbourethral glands

**Penis**. The penis consists of an attached *root*, a free *body,* and an enlarged tip, the *glans penis*. The penis is specialized with three columns of erectile tissue to become engorged with blood for insertion into the vagina during coitus. The urethra also passes through the penis as a conduit for urine. Erection of the penis depends on a surplus of blood entering the arteries of the penis as compared to the volume exiting through venous drainage. This is stimulated by parasympathetic innervation. Ejaculation, expulsion of semen through the urethra, is a result of sympathetic innervation.

**You Need to Know** 

**Erectile Tissues of the Penis:**

- **Corpora cavernosa penis:** paired, dorsal erectile tissue
- **Corpus spongiosum penis:** vental erectile tissue surrounding the urethra

## Female Reproductive System

The **primary female sex organs** are the ovaries. Female **secondary sex organs** develop prenatally and during puberty they mature and become functional under the influence of estrogens secreted by the ovaries. The structures of the female reproductive system (Figure 23-2) are described below.

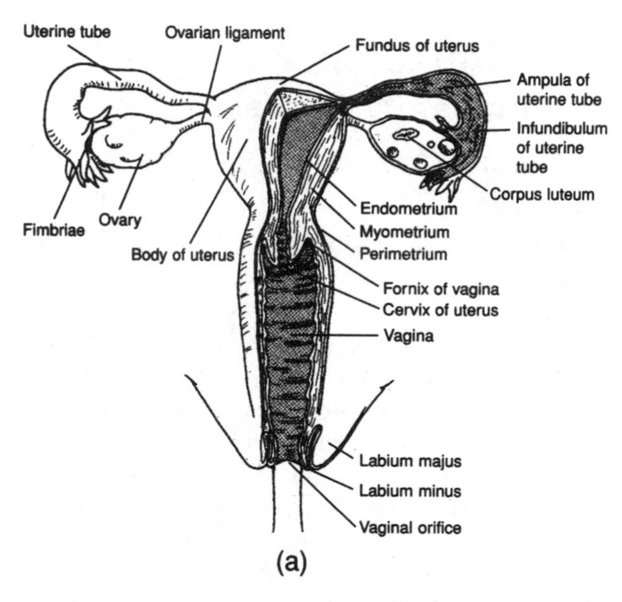


**Figure 23-2.** The female reproductive system. (a) anterior view

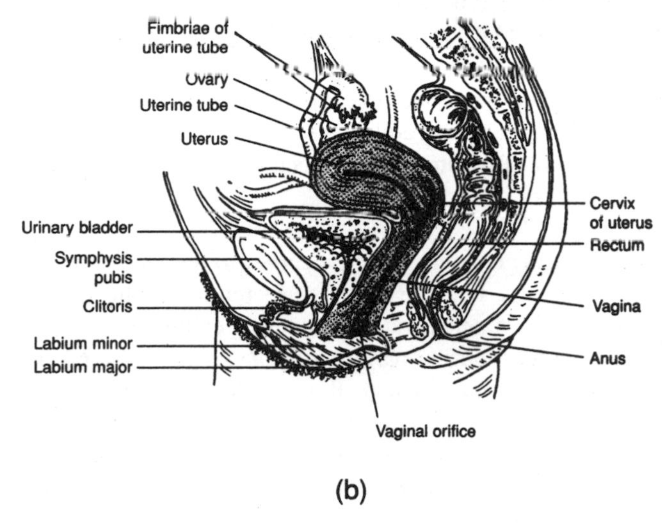


**Figure 23-2.** The female reproductive system. (b) sagittal view

**Ovaries.** The ovaries are located in the upper pelvic cavity, one on each side of the uterus. In the outer region of each ovary are tiny masses of cells, called **primary follicles,** each of which contains an immature egg. A group of follicles begin to develop at the beginning of a 28 day ovarian cycle; one reaches full development. The mature ovarian follicle secretes estrogen, which develops the uterine endometrium. At the middle of the cycle, the mature ovarian follicle containing a nearly completely formed ovum bulges from the surface of the ovary and releases the ovum. This is called *ovulation*. After ovulation, the follicle cells form the **corpus luteum,** which secretes progesterone and estrogens which also act on the uterine endometrium.

**Uterus and uterine tubes.** The **uterine tubes** extend from the ovaries to the uterus. They function to convey the ova toward the uterus and as the site of fertilization and early development (cleavage and formation of the blastula). The **uterus** is the site of implantation of the fertilized zygote and embryonic/fetal development. The parts of the uterus (Figure 23-2) are the *fundus*, the *body*, and the *cervix*.

Three layers of the uterine wall are:

- **Perimetrium:** Outer layer, composed of peritoneum
- **Myometrium:** Thick smooth muscle layer
- **Endometrium:** Inner mucosal layer composed of two layers
    - **Stratum basale:** deep layer, highly vascular
    - **Stratum functionale:** superficial layer, shed during menstruation.

**Vagina.** The vagina extends from the cervix to the vaginal opening of the external genitalia. It conveys uterine secretions to outside the body, and receives the erect penis and semen during coitus. It also is the passageway for the fetus during parturition. The vaginal wall has longitudinal folds, **vaginal rugae,** which allow for distension of the vagina.

**External genitalia.** The female external genitalia include *the mons pubis*, *labia majora*, *labia minora*, *clitoris,* and *vaginal opening*. Many of the structures of the external genitalia are homologous between males and females.

| Female structure | Male homolog |
|---|---|
| • Labia majora | • Scrotum |
| • Labia minora | • Body of the penis |
| • Clitoris | • Glans penis |
| • Vestibular glands | • Bulbourethral glands |

**Mammary glands:** Mammary glands within the *breasts* are accessory reproductive organs that are specialized to produce milk after pregnancy. Mammary glands are specialized sweat glands. At puberty, ovarian hormones stimulate their development. During pregnancy further development takes place under the influence of progesterone and estrogens. After parturition, secretion of prolactin stimulates milk production.

Stimulation of the *nipple* and *areola* during nursing results in the release of oxytocin from the posterior pituitary. Oxytocin stimulates the ejection of milk.

## Female Hormonal Cycle

Under the control of the hypothalamus, the anterior pituitary and the ovaries secrete steroid hormones. Cyclic changes in these secretions regulate all female reproductive activities (Figure 23-3).

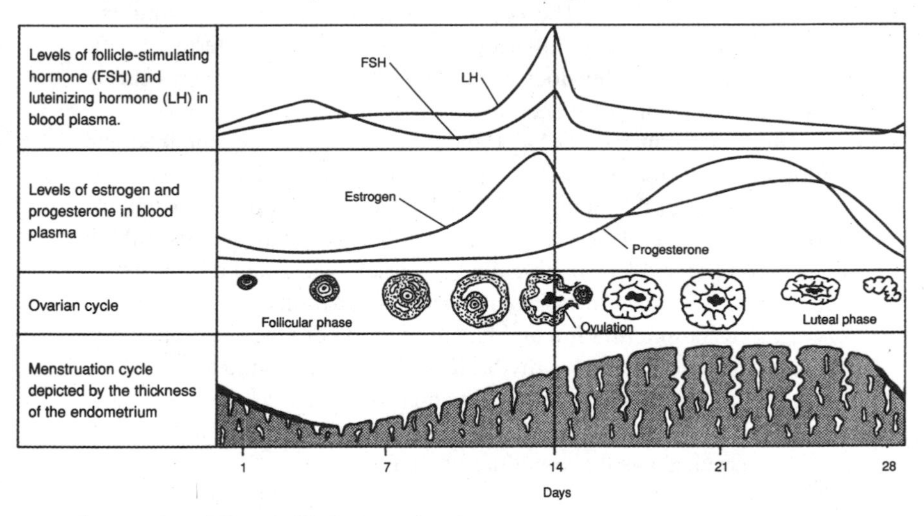


**Figure 23-3.** The menstural and ovarian cycle.

1. The hypothalamus releases luteinizing-releasing hormone (LRH); target organ is the anterior pituitary.
2. LRH stimulates secretion of follicle-stimulating hormone (FSH) and luteinizing hormone (LH), which stimulates follicle development in the ovaries (ovarian cycle).
3. The mature ovarian follicle secretes estrogens, which provoke a thickening of the endometrium (proliferative phase of the menstrual cycle, day 5 to day of ovulation).
4. After ovulation, the corpus luteum secretes progesterone and estrogens, which act to level off the endometrium and ready it for implantations (secretory phase of the menstrual cycle, day of ovulation to day 28).

5. With formation of the corpus albicans, the levels of progesterone and estogens drop, the endometrium decomposes, and a new menstrual cycle begins (ischemic phase of the menstrual cycle).

## Fertilization and Pregnancy

After being deposited in the vagina, spermatozoa move through the cervical canal and uterine cavity into a uterine tube to encounter the ovum on its way down from the ovary. **Fertilization** occurs in the distal one-third of the uterine tube. The **zygote** (fertilized egg) undergoes mitosis during its 3-day journey down the uterine tube to the uterine cavity. There the developing **blastocyst** remains free for another 3 days before it begins to implant in the endometrium.

Under the influence of hCG, the corpus luteum is maintained and continues to secrete progesterone and estrogens until the placenta is ready to take over this function. These hormones function to: sustain the endometrium, stimulate development of the mammary glands, inhibit release of FSH and LH (halting the menstrual cycle), and inhibit (progesterone) or stimulate (estrogens) uterine contractions.

**Labor** and **parturition** are the culmination of gestation. The onset of labor is denoted by the rhythmic and forceful contractions of the myometrium stimulated by oxytocin and prostaglandins. This is accompanied by cervical dialtion and discharge of blood-containing mucous from the cervical canal and out the vagina.

## Solved Problems

### True and False

____ 1. Meiosis is peculiar to the gonads. (**True**)

____ 2. Mammary glands are modified sebaceous glands. (**False**)

____ 3. Interstitial cells produce spermatozoa and secrete nutrients to developing spermatozoa within the testes. (**False**)

____ 4. The ovaries and uterus are the primary sex organs of the female. (**False**)

____ 5. Seminal vesicles, bulbourethral glands, and the prostate are all accessory glands of the male reproductive system. (**True**)

____ 6. The labia majora of the female genitalia are homologous to the scrotum in the male. (**True**)

____ 7. The secretory phase of menstruation is characterized by discharge of the menses. (**False**)

# Index